> Creativity flows when curiosity is stoked. Neil Blumenthal

TENDERING
WITH **INNOVATION**

For Micro, Small and Medium Industries

R.G. Chaudhari

INDIA · SINGAPORE · MALAYSIA

Notion Press

Old No. 38, New No. 6
McNichols Road, Chetpet
Chennai - 600 031

First Published by Notion Press 2020
Copyright © R.G. Chaudhari 2020
All Rights Reserved.

ISBN 978-1-64828-878-4

"Changes call for innovation, and innovation leads to **progress.**" Li Keqiang

TENDERING WITH INNOVATION

FOR MICRO, SMALL AND MEDIUM INDUSTRIES

R.G. CHAUDHARI

Success is fickle, but creativity is a gift. Tommy Shaw.

TENDERING WITH INNOVATION

FOR MICRO, SMALL AND MEDIUM INDUSTRIES

R.G. CHAUDHARI

HYDERABAD- INDIA

25 Jan 2020

CONTENTS

Topic No.	TOPIC	Page No.

PREFACE

This book is primarily meant for Micro, Small and Medium Enterprises (MSME) in manufacturing sector engaged in Job production and Batch production. Emphasis is on goods produced in factories with limited facilities and sub-contract support. Discussion on 'Tender' focuses on 'limited Tenders', 'Single Tenders' and 'Request for Quote'(RFQ) types of tenders. Here the customer is known.

Discussions are not limited to 'Tendering Process' alone but cover all other activities like manufacturing, Inspection, etc. relevant to objective of tendering.

First part of the book covers broad outline of tendering process; the second part deals with innovative ways of enhancing profit and third part lists the specimens.

It is universal fact that the principal objective of any enterprise is to **make profit** by selling products it manufactures (or providing services). But the problem is how to make profit. In tradition wisdom, attempt is made to cut cost of every element of expense, starting from procurement of raw materials to getting payment as early as possible. That is fine. But invariably, 'golden' chances of saving costs are missed due to strait-jacket thinking and over confidence.

To support this truth, this book is loaded with Real-life episodes or Case Studies from the author's personal experiences. They are described in first person narratives.

Therefore, while deliberating various aspects of tendering process, stress is laid on identifying scope for reducing expenses and wastages through innovative techniques because there appears to be no viable alternative for survival and growth without developing and improving products and processes. Also, this approach invariably improves quality, reliability and performance of the product(s).

An entrepreneur's life is not easy, especially in the early stages. He or she has to work round the clock to achieve daily and monthly targets. The result is a stressful life. Stress is, therefore, laid on the most effective solution for this problem viz. PLANNING: Start planning for the day: Take a piece of paper and make a list of jobs you need to complete for the day. Then prioritize them before you start your day. This helps you in properly orienting attention. They say: "IF YOU FAIL TO PLAN, YOU ARE PLANNING TO FAIL". Ensure that this habit of planning daily tasks becomes your second nature. There is no better way to lead a stress-free life.

Discussions in PART- I start with 'Enquiry' in <u>Activity Flow Chart</u> and culminate in 'Closure' of the order. PART-II demonstrates how innovations can increase productivity and profit dramatically.

ACTIVITY FLOW CHART

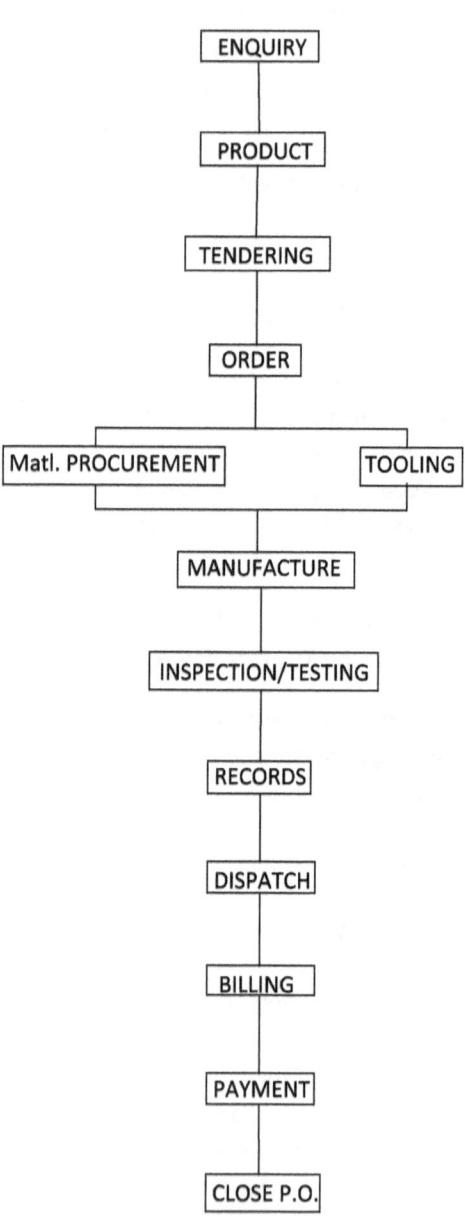

1. ENQUIRY/TENDER

1.0: AN ENQUIRY OR TENDER is an open request for bidding from suppliers for goods or services.

Requirement of all Government, Public sector and private sector undertakings are notified in print and electronic media.

1.1: TYPES OF TENDERS: There are more than a dozen types of tenders of which only the following three are discussed in this book.

(a) Limited or Selective Tender: In case of such tenders, only the pre-qualified or known bidders are invited to participate in the procurement process.

(b) Single Tender means sending the enquiry to a particular supplier where there is only one supplier or the product or process is of proprietary nature.

(c) Request For Quote (RFQ):

It is a 2-part bid consisting of (*i*) Techno-Commercial bid and (*ii*) Price bid.

The supplier has no scope to change anything expect perhaps price and sometimes delivery schedule.

It is a transparent system and used in e-Procurement system if the order is over a certain value.

1.2: TYPES OF BIDS: Bids are supplier's proposals against enquiries. There are two types of bids:

(1) Price Bid and (2) Techno-Commercial Bid.

(a) Price Bid includes price, taxes, levies, duties, freight charges, etc. (i.e. money matters)

(b) Techno- commercial Bid: The first part of this bid deals with technical aspects such as specification, drawings, inspection, testing, etc. and the commercial part elaborates General Conditions of Contract (GCC), Special Conditions of Contract (SCC) and Instructions to Bidders (ITB).

(c) Envelope Biding: In Government and Public sectors, more commonly used systems are (1) Single- Envelope Bidding and (2) Two- envelope Bidding.

In single- envelope bidding, the price bid and techno-commercial bid are submitted in the same sealed envelope.

In two- envelope bidding, price proposal and technical proposals are enclosed in two separate sealed envelopes and then both are put in a single large envelope. Tender No. and the date of tender opening; and Price Bid or Techno-Coml. marked in bold letters on the first two envelopes.

First the techno-commercial bid is opened to find whether the bidder meets all technical and commercial requirements. The price bids of only technically qualified bidders are opened subsequently at a later date.

2. PRODUCT

2.0 PRODUCT: It includes Product Group also in which various sizes of an item are covered by a single Specification or Product Standard.

An enquiry is accompanied by the drawings, specifications and other documents pertaining to terms and conditions of supply of the product(s).

Before submitting your tender, you should have studied all aspects of the product (group). The primary document for this study is specification, also called Product/ Plant Standard. It describes:-

- Function/Scope of the product,
- Sizes and Tolerances.
- Material specification.
- Dimensioned drawings of parts and assemblies.
- Technical Requirements like material of construction (MoC), Manufacturing process, Working conditions (Temperature & Pressure), etc.
- Types of Test: Hydro-test, Liquid Penetration (LP) test, Material test, etc.
- Preservation and packing: Rust prevention coating of ferrous parts, Polishing of SS Parts, protection of threads by caps and plugs, etc.
- Marking Details: Logo and Unique ID No. of supplier, Thread sizes, Material grade, Working and test pressures, etc.
- Many a time, cross references are given for other tests such as PMI (Positive Material Identification), MPI (Magnetic Particle Inspection), LPT (Liquid Penetration Test), WPS (Welding Procedure Specification), Radiography, etc. Facilities/equipment for conducting such tests may not be available in MSMEs and charges of NAB* Labs will be prohibitory. Hence, you should look for such clauses with hawk's eye and never forget to include the cost of such test(s) in price bid. Ascertain the costs of such tests from the Laboratory.

* NAB Lab: National Accreditation Board for Testing and Calibration Laboratories

2.1: PRODUCT REQUIREMENTS:

What are the Product requirements, as expressed through Specification/Product Standards?

Briefly they are described below.

(For further details, refer to chaper-3, 'Tendering').

2.12: Dimensions:

a) Max./Min.; Medium, Major or critical.

b) Fits and tolerances, cumulative or non-cumulative.

c) Measuring Instruments: Ranges, accuracy, validity of calibration, etc.

d) Marking devices: Surface Plate, V-blocks, height gauge, etc.

2.13: Raw Materials:

a) Specification::National, International or Company Standards.

b) Grades and Conditions, e.g. hardened & tempered.

c) Chemical composition and Mechanical properties.

d) Shapes: Bar Stock Cross section: Round, Square, Hexagonal, Rectangular; Sheets, Plates, etc.

e) Castings: Sand/Die/Investment castings.

f) Forgings: Rough/Die forged

g) Pipes: ERW, Seamless - Nominal Size, Schedule, etc.

2.14: Bought-out Items:

a) Pipe Fittings: Tees, Elbows, Unions, Couplings, Reducers, etc.

b) Fasteners: Nuts, Bolts, Circlips, Clamps, Cotter pins, Dowels, Rivets, etc.

c) Springs: CS/SS spring wire, Load Vs Deflection characteristics,

d) Sealing Items: Gaskets, 'O' rings, Washers (Al.), etc.

e) End Closures: Caps and Plugs in metal and plastic.

f) Packing Materials: Boxes, Cartons, Thermocol, Paper shredding, Adhesives etc.

2.2: PROCESS REQUIREMENT:
(Facilities and capacities: In-house/Sub-contract).

2.21: Material preparation:

a) Cutting/Shearing, Blanking, Gas cutting, etc.

b) Castings and Forgings: Patterns and Core boxes; Forging dies; Fettling, Shot/ grit blasting.

c) De-burring: Abrasive products.

d) Transfer of attestation marks (IBR*, Lloyd's, PDIL, etc.) * Indian Boilers Regulations

e) Hard punching of ID mark, size, grade, etc.

2.22: Machining:

a) Center Lathes: **Machining dimensions and attachments**

b) Milling: Universal, vertical or horizontal milling machine and attachments.

c) Drilling: Bench, pillar, radial or gang drilling machine

d) Slotting: Internal or external.

e) Grinding: Surface, Cylindrical (internal/external), etc.

f) CNC Machining.

2.23: Press Operations:

Machines* required for: Shearing, Blanking, Notching, Forming, Drawing, Bending, etc.

(* **They include: Shearing m/c,** Screw Press, Hydraulic Press, Press Brake, Power Press, etc).

2.24: Finishing Operations:

a) De-burring: Bench Grinder, Straight/angle grinder, Belt-cum-disc sander, Shot-blasting, etc.

b) Polishing, lapping and buffing,

c) Surface Coating: Primer and Finish Paints, Epoxy, Plating, Black-oxidizing, Anodizing, etc.

2.25: Assembly: By screwing, nut-bolts, riveting, brazing, welding, etc.

2.3: PROCESS CONTROL:

2.31: Manufacturing Quality Assurance Plan (QAP):

a) In process Measurements.

b) Check-Points.

c) Customer hold-Points.

d) Final Inspection.

Full details of QAP are given in chapter '**8: INSPECTION AND TESTING**'.

2.32: <u>Quality Control:</u>

a) Instruments and gauges with valid calibration certificates

b) Jigs, Fixtures and templates.

c) Identification, segregation and disposal of non- conforming parts.

2.33: <u>Tests:</u>

a) Applications/Field conditions:

- Hydraulic Pressure Test.

- Performance Test.

- LP Test.

- Radiography

b) Material tests:

- Chemical Analysis

- Mechanical Properties.

- Radiography of finished goods.

- Material Positive Identification at NAB Lab

- Magnetic Particle Test.

- Ultra-sonic Test

at NAB Lab

2.34: <u>Attestation and Marking:</u>

a) Third Party Attestation.

b) Customer's attestation.

c) Applicable Specification

d) Marking:

- Product/Item code.

- Material grade.

- Pressure Rating.

- Thread sizes

- Tag No.

2.35: <u>Preservation:</u>

a) Protective coat of paint, Electro-plating, anodizing, etc.

b) Black-oxidizing.

c) End closures (Plugs and caps).

d) Thread Protection-Plastic Plugs and Caps.

2.36: <u>Packing:</u>

a) Wooden boxes and crates

b) 'Thermocol' sheets, boxes.

c) Paper shredding.

d) Cartons.

e) Metal tags bearing item codes.

f) Packing list.

g) Consignee Address, etc.

The lists given above are check lists. Pick up only those items which are relevant to your production needs.

2.4: <u>PRODUCT BACKGROUND:</u>

- Was the product quoted earlier?

- Was a prototype developed and tested earlier but not quoted?

If the product was quoted or its prototype developed earlier, details of all parameters such as costs of materials, operations, testing, etc are available on your records. All that you have to do now is to revise item-wise rates.

- Is the product to be quoted for the first time?

2.41: <u>First time quote:</u>

If the product is to be quoted for the first time, make lists of requirements of:

- Raw materials: Ferrous, Non-ferrous, non-metallic, shapes, sizes, etc.

- Physical facilities like machine tools, cutting tools and jigs & fixture to convert raw material into finished product.

- Quality Control (QC) instruments with valid calibration certificates.

- Testing Equipment.

- List of operations to be carried out in your factory and at sub- contractors' works.
- List of Bought-out items with quantity and rate of each item.

Such lists are *sine qua non* for preparation of a structured quote.

N.B.

While going through this book, you will come across names of "OIL RIGS' products that you might not have heard of earlier. You will find brief description and drawings of such products in Author's book titled "TITBITS AND BOUTS OF CREATIVITY". It is available at Amazon, Flipkart, Notion Press Store, etc.

3. TENDERING

3.0: TENDERING:

Buyers publish tenders and suppliers respond to their requests by submitting proposals.

Acceptance of supplier's proposals typically results in a contract, commonly called Purchase Order (P.O.). **This process is called 'Tendering'.** It is the heart of your endeavor to secure an order with handsome profit.

3.1: AD HOC METHOD

It is observed that common trend in preparing a price bid is to add up **estimated** costs of:

(a) Raw material, (b) conversion, (c) ad hoc value of overhead expenses and (d) certain percentage of costs as profit. Some suppliers' thumb rule is to increase or reduce certain percentage from the last price of previous successful competitor.

Such ad hoc practices of calculating selling price results in low margin or even loss.

Instead of depending on hunches, take knowledge-driven decisions. (Refer: Art. 3.4).

3.2: PRE-TENDERING ACTIVITIES:

An enquiry is accompanied by a set of documents like specification, drawings, GCC, STC, ITB, etc. You might have received such documents from your principal customer with first enquiry and preserved their copies. Also, you may go to internet and download documents related to e-Procurement Process of ONGC, IOL, NTPC, NLC, BHEL, RELIENCE, your principal customer(s), etc. (e- Procurement system of NLC is completely computerized and worth going through)

Bidders are advised to get acquainted themselves with various requirements of buyers by exercising abundant care in submitting their offers.

After all, your aim is to WIN CONTRACTS AT THE RIGHT PRICE

You should scrutinize the conditions of contract very carefully and memorize the following clauses in particular:

- Inspection and tests.
- Packing & types.
- Freight charges
- Delivery conditions.
- Penalty for delayed delivery.
- Guarantee.

3.3: THE LATEST TREND IS TO GO FOR E-PROCUREMENT PROCESS AND REVERSE AUCTIONING.

(a) e-Procurement:

For participation in e-Procurement, minimum requirements are:

(1) PC connected to internet.

(2) Registration with portal.

(3) Class-2 or class-3 digital signature certificate.

(b) e-Procurement portal of Government of India:

For details of e-Procurement system, go to this portal and download the tender schedule, Special Instructions to Bidders (SITB), etc. and then submit the bids online:

For conventional or e-procurement system, discussion on the following topics is common.

3.4: PART-I : PRICE BID:

'Use your business acumen in pricing but only after thorough analysis of costing structure.'

Before analyzing cost structure, study the Requirements of Product, Process, Quality, Preservation, Packing, etc. as discussed in the previous chapter.

Most important task of tendering is the preparation of PRICE BID which should be based on a **structured system and real time costs.**

Broadly, expenses incurred in running a factory fall under two types:

(1) Direct Expenses that directly contribute to the manufacture of products.

(2) Indirect Direct Expenses: Expenses that do not contribute directly to production.

3.41: Direct Expenses/Manufacturing cost:

a) Raw materials (RM) as quoted by the suppliers (Add min. 2% contingency).

b) Conversion of RM into finished goods (Operation cost)

c) Job work Charges paid.

d) Wages, OT wages, Bonus, Incentive and Rewards.

e) Bought-out items like Nuts, bolts, washers, steel Balls, Lock nuts, special paints, "O' rings etc.

f) Consumables: (Apportioned cost of Common tools, Lubricants, Gases, Electrodes, Chemicals, etc).

g) Finishing operations like polishing, buffing, painting, electro-plating, black-oxidizing, etc.

h) Apportioned cost of development, if any.

i) Electricity and water.

Manufacturing cost: By adding these costs, you arrive at **manufacturing cost.**

3.42: Indirect Expenses and Overheads:

a) Remuneration to CEO (Fixed and/or certain percent of profit)

b) Salaries and wages of Office Staff like Computer Operator, Accountant, Clerks, Watchman, etc.

c) Employee Welfare: Refreshment, Uniforms, Safety Shoes, Gifts/Sweets, etc.

d) Business Promotions: Ads, Broachers, Gifts, Hospitality, etc.

e) Inspection : Apportioned cost of calibration of instruments and gauges at NAB Lab, Pressure Test, L.P. test, Material Tests (Chemical analyses and Mechanical properties), attestation under IBR or by Lloyd's, etc

f) Repairs and Maintenance of assets: Machine tools, equipment, computer, building, etc

g) Interest to Bank and bank charges

h) Vehicles up-keep.

I) Conveyance

j) Rates and Taxes

k) Tour and Travel.

l) Transportation/freight charges. (Contact Transporter for rates)

m) Transit Insurance for out-station deliveries (Contact Insurance Company for rates).

n) Tax Consultancy fee.

o) Audit fee.

p) Depreciation on assets: Land and Building, Patterns, Mould and Dies, Machinery and Equipment, Vehicles, Electrical fittings, Office Equipment, Computer/Electronic devices and so on.

q) General Expenses: Purchases of petty items like Ropes, 'Sutli', Chalk, Brooms, M/C brushes, Cotton waste, Kerosene, Putty, etc.

Overhead Expenses: By adding the above costs, you arrive at Overheads.

3.43: Factory cost: To manufacturing cost, add overheads to get **Factory Cost.**

3.44: Selling Price: Based on degree of competition and your business **policy***, decide as to what percentage of factory cost should be added as profit to arrive at **Selling Price.**

(*Please note: Price is a policy and cost is a fact. One may quote lower than cost to 'kill' the competition or add 500% margin to the factory cost if the product is of proprietary nature). Instead of resorting to such extreme steps, the best policy is to keep the margin at certain percentages.

3.45: Duties, levies and taxes: Applicable values are added to Selling Price in deciding **'L1' offer.**

3.46: NOTES:

a) **Apportioned Cost:** You may consult your Auditor for calculating 'Apportioned costs' and 'Factory and Administrative Overheads'. Thumb rule is to take an average of past three years and express it as a percentage of past three annual turnovers.

You can calculate the apportioned cost and Overhead from the figures given in your 'Profit and Loss Accounts' of the past 2 or 3 years.

b) **P & F Charges:** Packing and forwarding (P & F) charges may be quoted as 2% to 3% of basic price.

Annexure-1 shows a sample list of Direct and Indirect expenses.

c) **Costing sheets:** Item-wise/Part-wise cost structure for a product once prepared, must be preserved for future reference. It may be a resister or a folder in computer. It should be up-dated periodically.

Two examples of costing sheet are placed at Annexure-2.

d) **Quantity Discount:** As a rule, avoid offering discount on prices. However, to win the game of 'averages' you may offer quantity discount on some condition like: "If all items of enquiry are ordered together....." or "If the validity of contact is extended for 2 years.

If your prices are lowest, politely reply saying "We have offered rock-bottom prices; hence regret not to find any scope to reduce prices or to offer discount". If your product(s) is (are) of monopoly nature with high profit margin(s), you may offer 2 to 3% discount 'to maintain our cordial business relations'.

You may compromise on price but never on quality.

3.5: PART-II: TECHNO-COMMERCIAL BID:

3.51: Technical Bid: The first part of this bid deals with technical aspects like specification and tests. It is outlined above under Product Requirements, Process Requirement and Quality Control.

3.52: Commercial bid:

As stated herein above, every tender is accompanied by a set of documents, of which GCC, SCC and ITB deal with commercial aspects of tenders. It is imperative that the important features of these documents given below are scrutinized very carefully and **memorized**.

a) **Inspection and Tests:** Inspecting agency, stages of inspection, types of tests like measurements, performance test, Hydro-test, etc. in compliance with certain specification. (Normally, Type Test is not in the scope of MSME).

b) **Preservation:** Surface coating is essential to protect the ferrous material from rusting, tarnishing and loosing luster (of polished and buffed surfaces). Coating of grease and oil, Painting, Black -Oxidizing and Electro-plating are the common methods of preservation. However, utmost attention should be paid if 2-part Epoxy painting is specified.

3.53: Case Study -1. Painting.

This incident was reported to the author by his associate.

The job was a sheet metal container of the size about: 2 M x 1.2 M x 1 M deep.

The specification prescribed painting process as: Shot-blast outside surface clean to SA2½ before painting. Apply rust inhibitor. Primer: zinc silicate ~ 75 Micron. Finish Coat: 2 coats of Repack, High- build Polyurethane 125 Micron.

The Third Party Inspector (TPI) checked the thickness of the paint and found it to be about 100 µ. On scrapping the paint at one corner, he found no coat of primer. Blame game followed.

Cause: Supervisor was not aware of Epoxy painting and the owner has failed to study the procedure and to educate the supervisor.

Result: Loss of over Rs. 12,000/- due to loss of very costly paints and many man-hours in rework.

c) **Packing:** Packing should withstand transportation by Rail, Road, Air or Sea as specified. Types of packing vary from consignment to consignment. It is elaborated in Conditions of Contract.

d) **Delivery term:** Watch out for the clause: "Delivery shall be FOR destination", where FOR stands for Freight On Rail or Road.

Destination may be buyer's stores in your town or at a Sea Port 1500 KM away from your works.

Accept FOR clause after ascertaining its cost from your transporter by furnishing weight/volume. (He calculates freight charges by multiplying weight or volume by distance).

Do not forget to add cost of transporting goods from your works to transporter's booking office or to the local Stores of the customer.

e) **Delivery Schedule: This clause has serious implication in receiving your payment.**

You are advised to quote time for delivery in weeks. For FOR destination/Project site, there are two ways of interpretation:

(i) Delivery at customer's approved transporter is reckoned as date of delivery. (No problem)

(ii) Delivery at Project Site and 'Certificate of Material Received' (CMR). This is a tricky issue.

  - Transporter may take several days to load your goods till he gets truck-full of load to destination.

- Transit time is also unpredictable.

- Having received the material, Project Chief takes his own sweet time {up to 4 to 6 weeks) to issue CMR (Certificate of Material Receipt). Unless his certificate is submitted, your bill will not be taken for processing. Therefore, insist on transporter to follow up for CMR on daily basis.

f) **Insurance Charges:** For calculating F.O.R. destination cost, seller adds insurance charges to other cost elements. Typically, it is quoted as percentage of basic price. However, it is better to approach your Insurance Company with value of consignment, destination & mode of transport.

But never forget to take out a transit Insurance policy.

g) **Penalty Clause:** Delivery is the essence of all types of procurement systems. If the delivery is delayed beyond the stipulated date, your customer recovers penalty @ 0.5% or 1% per week from your bill, subject to maximum 5% or 10%, respectively

3.54: Case Study -2: Penalty Clause.

(Note: Product name and order value changed in the example).

"In the 12th year of operation of my SSI Unit, I had received an order for supply of three sizes of Thermowell worth Rs.2,60,000/-(Basic price).

GCC specified penalty @ 0.5% per week. The delivery was delayed by 2 weeks, Finance dept. recovered penalty Rs.5,200/- @ 1/2% for 2 weeks from my bill. On protesting, Accountant showed me the Special Instructions to Bidders (SITB) sheet signed by me. In SITB, penalty was mentioned as 1% per week.

I relied on my memory and assumed penalty clause in GCC to be unalterable. Had I read this clause carefully in SITB, I could have avoided such awkward situation.

Consider other implications of paying penalty.

- I have lost the penalty amount forever.
- I was borrowing working capital from a Bank and paid interest on penalty amount perpetually.
- I lost an opportunity to invest the penalty amount, say, in fixed deposit and to earn interest."

4. ORDER

The sweet reward of your efforts is the Purchase Order (PO).
PO is a contract between you and your customer.

4.1: ORDER ACCEPTANCE AND CONTRACT REVIEW:

Having secured an order, your first job is to check each and every clause/entry in the PO in comparison with clauses in the documents received with enquiry/RFQ. Many a time, particularly for Direct Dispatches (and Sub-contract orders), entries under P & F and Freight will be vague like 'As applicable' or 'Extra'.

Also, the delivery schedule may be shorter than cycle time. This results in your paying penalty.

List out all non-conformities and make a written request for amendment.

If the P.O. is in order in all respects, acknowledge its receipt and confirm delivery schedule, giving your unique Id No., PO No. and date, etc.

A specimen of Order Acknowledgement is given in Annexure-3.

4.2: ORDER BOOK:

On receipt of an order, assign a unique Identification number to each item in the PO.

This unique ID No. starts with first item in the first PO and continues uninterrupted. It is stamped and/or written on many documents and hard- punched on product.

For reference to PO containing more than one product, Unique ID No. of the first Item in the list is used.

Next, enter the particulars in the Order Book with the following 14 Nos. of column headings:

Date of receipt of PO, Sl. No., Unique ID No., Description, Item or Product Code, Quantity, Rate, Order Value, Delivery Date, Inspecting agency/date of Inspection, Delivery Challan No./Date, Invoice No./Date, Payment Received date and Remarks (Project, Penalty, etc.)

This requires an Excel sheet or two facing pages in the Register (Order Book).

Order Book is one of the most important records. After searching his computerized record, your customer may refer to a 5-year or 10- year old PO placed on you. In such events, your Order Book comes handy for quick response.

5. MATERIAL PROCUREMENT

5.1: PROCUREMENT: While preparing PRICE BID, you have made lists of Raw Materials, Consumables, etc. with quantity, rate and total requirement against each item of enquiry.

If you do not have a running rate contract, invite quotations from at least two reliable suppliers spelling out your terms and conditions:

a) Whether prices are on door delivery basis.

b) Delivery schedule for procurement from out-station supplier(s).

c) Whether Mill Certificate or Material Test Certificate (MTC) from a reputed Testing Lab will be provided and whether Lab charges are inclusive.

d) Payment term (say, within 2 weeks from the date of receipt and acceptance of material).

e) In case of castings, it should be made very clear that castings should be free from foundry defects like 'cold shut', sand inclusion, shift in core halves, blow holes and so on.

f) P & F charges (preferably in percentage of order value)

g) Taxes and Duties.

Purchase Order: The next step in procurement is to place a Purchase Order on the supplier whose offer is closest to your purchasing policy, namely:

"Right Price, Right Quality, Right quantity and On-time Delivery".

Invariably, MICRO and SMALL enterprises purchase the following items from local dealers without placing formal purchase order:

- Carbon steel, Alloy Steel and Stainless Steel bar stocks, plates, sheets, flats, etc.
- High tensile fasteners, Gasket sheets, Acrylic tubes and sheets, 'O' rings, etc.

Such items should also be procured by keeping your Purchasing Policy in mind.

On receipt of material, your first job is to send the material for Chemical and Mechanical tests to an NAB lab. It should be ensured that Mechanical Test was witnessed by the customer's inspector.

The first operation on the raw material starts only after it passes the said tests.

Although some raw material and bought-out items are 'cheap', you should avoid building up of inventory. Inventory is like termites nibbling into your profit. The best policy is to aim at 'zero' inventory. This is one of the most recommended cost cutting tools.

6. TOOLING

6.1: LISTS OF TOOLS:

Tools are generally categorized as (1) Hand Tools, (2) Portable Power Tools, (3) Machine Tools and (4) Computer-aided Tools.

In the previous article, we have discussed ways and means of saving costs in the procurement of raw materials and consumables. The process of procuring tools and tackles is no different.

In machine shops, production involves cutting, shaping, drilling, finishing and assembly.

Based on your experience and day-to-day needs, you might have purchased Hand Tools, Power Tools, Machine Tools, Jigs, Fixtures, Dies, etc. to carry out these operations.

Lists of Tools commonly used in Mechanical Industries are placed at Annexure-4 for reference.

It is assumed that following occasionally required operations are off-loaded:

Cylindrical grinding, Surface grinding, Honing, Broaching, Slotting, Shot blasting, Nibbling, Rolling, Folding, etc. which are rarely required and investment in machinery to carry out such operations in your shop is not justified

In order to reduce investment in fixed assets, you might have developed some tiny 'industries' each run by one or two skilled artisans possessing 1 or 2 machines including CNC lathe, for taking up your machining jobs on rate contract basis.

"What can be done elsewhere should not be done here, i.e. in your shop" is one of the best strategies to reduce production cost significantly. However, you should have minimum facilities to get approval as vendor by your customers.

Selection of cutting Tools: Depending on the properties of materials to be cut and shaped, it is your job to decide which of the following material(s) is (are) appropriate for the purpose:

- Carbon tool steel, - High Speed Steel (HSS), - Carbide Tips, -Ceramic Tools and Diamond tools for: Cutting, Chipping, Threading, Boring, Shaping, Parting and so on.

Example: Let us assume that a Turner is provided carbon steel tool for shaping a rod of EN9 because it was cheap. Now, observe the number of trips he makes to pedestal grinder for re-sharpening the tool, and frequency of tool replacement. Consequently, you incur more expenses rather than saving on production cost.

The backbone of your Price Bid is the money you save on every element of cost.

In saving production cost and maintaining quality, dies, jigs and fixtures play a very important role. There is a vast scope to reduce cost of dies by manufacturing them in the shop.

Example-1: For about 300 numbers of plastic caps for protecting M33-M threads of thermowells, Plastic Moulder charges, say, Rs. 1200/- for the die and Rs.3/- per piece. He retains the die. For future requirements, you pay the price he demands.

The same is the case with other sizes of caps, plugs and 'O' Rings of VITON.

As a cost saving strategy, buy one mouding die from a professional, study the 'technology' of manufacturing plastic moulding dies and make them in your shop.

Example-2: For drilling 4 holes in a small batch of flanges, positions of holes are marked by hand. This method does not assure the accuracy, leading to some rejection or re-work. It is economical to use a drilling jig plate. A single plate with holes on different PDCs (Pitch Dia. Circles) saves the cost of raw material.

6.2: CASE STUDY -3: Machining of Square body in 3-Jaw Chuck.

(Refer to Annexure-2). "160 numbers cubical end blocks for Liquid Level Indicator were to be faced, drilled, bored and threaded on one face of each cube. Machining on lathe with 4-jaw chuck produced 6 or 7 numbers per shift. Delivery was critical. Sub-contractors declined to take up the job on their CNC lathe, saying that round jobs only can be loaded on CNC lathe. (Each side of a block is 40 mm).

I saw an opportunity to dismantle this mind-set.

Solution: Here is the solution. (See drawing on the right)

Notes:

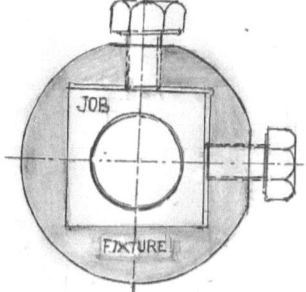

1. a stopper flat 6x12x50 is welded at the back.
2. Length of fixture is about 35 mm.

3. Similar fixtures were made and used by our Turner on centre lathe for machining bodies of Adj. Orifices and jobs with rectangular cross sections.

4. Productivity on CNC lathe and conventional lathe had gone up by about 8 times and 3 times, respectively.

The author says: What is new to me may not be new to others but as long as I believe it is novel and useful idea, I go ahead.

This was the simplest idea. The author had made tens of innovative fixtures. Some of them are included in his book titled 'TITBITS AND BOUTS OF CREATIVITY, referred to as **BOOK** [2].

7. MANUFACTURE

In the process of manufacture, there are some activities which run sequentially and some, concurrently. As an example of sequential operations, take machining of a pipe union. Operations are: cutting a round bar followed by turning, drilling, threading, chamfering and finishing. You cannot change the sequence.

All planning activities like preparing cutting plans, procurement of consumables, ordering 'Item Code' tags, etc. can be done concurrently.

7.1: PROGRESS PLANNING:

Your first job in manufacturing process is to make a simple plan showing the sequence of operations for each part, name of the work station and estimated time of operations.

- **Cutting:** Power hacksaw, Cut-Off M/C, slitting saw, gas cutting, shearing or parting on lathe m/c, etc. (Choose one or two, as applicable)
- **Machining:** Turning, milling, drilling, Threading, etc.
- **Grinding:** Cylindrical grinding (Internal, external or both), surface grinding.
- **Finishing:** Sanding, polishing, Oxidizing, Case hardening, etc.
- **Assembly:** A multi-part product requires assembly of parts either by screwing or by using fasteners, riveting or brazing or by welding or combination of two or more operations.

(There are hundreds of operations that are carried out in multi-product factories. However, only few of them may be relevant to your products.

- Ensure that all required operations are entered into your Process Sheet, noting the name or number of the special tool, jig or fixture to be used.

Once a 'Process Sheet' is made for a product, it should be preserved for future use, and reviewed from time to time.

7.2: STANDARD TIME:

In manufacture, raw material is converted into finished product by using various types of tools and skills. To evaluate the efficiency of men and machines, you need some yardstick. That yardstick is called Standard Time (for each operation.)

Standard Time means 'the time required by an average skilled worker at a normal pace, to execute a specified task using a prescribed method (technology)'.

Setting up Standard Times is a job of an Industrial Engineer. MSMEs may not afford to employ him on full time basis. However, they may seek the help of National Productivity Council (NPC) in developing standard times for some major/critical jobs.

For internal use, you can calculate standard time per unit by observing number of units 'N' produced in 'T' man-hours. Standard Time = N÷T. N and T should be the averages of at least 3 readings taken in different shifts with different workers.

In big organizations, Methods or Technology department provides 'standard times' to Sub-contract cell or Accounts Dept. who have 'Hourly labour Rates 'for all skills. The cost of any operation is then calculated by multiplying standard time with rate.

MSMEs: In India, they are some systems of employing workers in manufacturing industries governed by Factories Act, Minimum Wage Act, Bonus Act, etc. Normally, MSMEs prefer to keep the workforce on Nominal Muster Roll and pay (minimum) wages on monthly basis as per Minimum Wage Act, and Bonus on yearly basis (Min. bonus is 1/12 of wages earned in a financial year). In such cases, **CALCULATION OF DIRECT MANUFACTURING LABOUR COSTS** for small businesses can be done as follows:

Step -1. Pull out Attendance Register (or Clock-in System Record) and find out how many hours each worked on production process from the beginning to end. Add up the hours each worker worked to arrive at the total numbers of hours. For example, say you had 5 direct workers and each of them spent 200 hours on the given job. Total numbers of hours will be 200 x 5 = 1000.

Step-2. Open your payment Register and find out the total wages paid to direct workers for the relevant month. Example: Say that you had paid Rs. 9,600/- plus apportioned bonus @ 15% , that is, Rs. 120, making it 9,600 + 120 = 9,720/- per month per worker. Total amount paid to 5 workers will be Rs.9,720 x 5 = 48,600 for the month of 30 days or 30 x 8 = 240 hours per worker. Total labour hours for 5 workers will be 240 x 5 = 1200 hours.

<u>**Amount of actual rate per direct labour hour = Total Amount paid ÷ Total labour hours**</u>

$$= 48600 + 1200 = 40.50$$

If you are off-loading operations, find the rates from your sub-contractor and compare them with rates calculated by you. Your rates will be the basis for negotiations.

7.3: RANDOM THOUGHTS ON INSPECTION

Having taken all care in manufacturing a top quality product, you may be wondering as to why your customer does not accept the goods as offered by you without inspection. You may site examples of branded goods. Branded goods are manufactured on large scale. The process, process controls and inspection procedure are certified by National Bureau of Standards.

Goods manufactured in your factory are against specified Standards in measured quantities.

Your customer or customer's customer does not buy products that you can manufacture. The product is acceptable to him/her only if it fulfils his/her requirements, that is, if it is manufactured as specified in his Product Standard and is in conformity with terms and conditions. The Product Standard specifies material grade, ways and means of controlling process of manufacturing, quality level through the use of Inspection and Testing tools

Secondly, If yours is an ISO certified Company, it means your organization has integrated the six quality policies and procedures specified by ISO, namely:

(1) Document Control

(2) Control of Quality records

(3) Control of Non-conforming products

(4) Corrective Action

(5) Preventive Action and

(6) Internal Audit.

Discussion on the topic of ISO is not in the scope of this manual.

8. INSPECTION AND TESTING

Inspection and Testing are the means of determining the quality level of the product. They are essential tools to control quality, identify causes of defects and reduce or eliminate rejections.

Quality cannot be inspected. It is to be built in the product.

Inspection of the manufactured goods starts soon after receipt of material and ends in Packing.

8.1: QUALITY ASSURANCE AND QUALITY CONTROL.

Let us first understand the difference between Quality Assurance (QA) and Quality Control (QC)?

Briefly, QA is a written procedure that will assure full compliance with all contract requirements. In other words, its goal is to prevent defects in manufactured product(s) and provides assurance that the required quality will be achieved. QA relies on specification/Standard of the product.

QC focuses on identifying and fixing defects. QC follows QA.

In the documents accompanying purchase order, your customer asks you to submit a 'Manufacturing Quality Assurance Plan' (QAP) in a prescribed format. Your customer also provides 'Guidelines to Vendor for Preparing the QAP'. You are advised to read it carefully before preparing the QAP. Based on Process Sheet and Spec., you prepare the QAP and submit it for approval.

A specimen of QAP is shown in the ANNEXURE-5.

In spite of taking almost care in preparing a QAP, don't expect that it will be approved in 'to to'.. Quality Engineer will dig deep into the specification; and taking advantage of some unwritten or vague clause, adds his comment(s) to QAP.

Example-1: Specification for Orifice Plates gives MoC as SS304. I machined the product from SS PLATE because the name of product contains the word 'plate'. In QAP, QA Engineer changed 'Plate' to 'Round Bar Stock'. The orifice plates were already machined and kept ready for final inspection. The blame game ensued. Finally, the matter was resolved by Design Head: Product Standard was revised by changing MoC to <u>ASTM-A276 TP</u> 304; and this batch of orifice plates was accepted 'as **one time exception'**

Example-2: This pertains to R1 Threads (i.e. Taper threads as per IS: 554). In the QAP of Thermowell under the column heading 'Reference Document', I wrote ISO 7 (or ISO 7-1). Q.A. Engineer promptly changed it to IS: 554 in QAP.

The opening paragraph of the Spec. for IS: 554 reads:

"This Indian Specification is <u>identical</u> with ISO 7-1.... Dimensions, Tolerances and designation issued by I.S.O. were <u>adopted</u> by B.I.S ......").

Then, why had the QA Engineer changed ISO:7 to IS: 554? Because, the Co's specification says IS:554.

8.2: IN-HOUSE INSPECTION:

First, the manufacturer conducts 100% inspection in respect of Material, Dimensions, Hydro-test, Surface finish, etc. as detailed hereunder

Let us now recall the events from procurement of raw materials to submission of QAP.

1. Quality of raw material was tested in NAB Lab and MTC(s) was (were) obtained. These tests include:

   (a) Chemical composition and Mechanical strength.

   (b) Positive Material Identification, if required.

   (c) Magnetic Particle Test, if called for.

   (d) Liquid Penetration Test, particularly for Castings, forgings and welded parts

2. The process of converting raw material into finished product(s) started with cutting operation: i.e. 'material preparation'

3. After completion of various machining operations, the work on 'In-process Inspection' starts.

8.21: Dimension Inspection Report (DIR)

In case of machined components, inspection starts with measurement of linear and circular dimensions, using calibrated instruments. Threads are checked by 'GO' and 'NOGO' gauges.

On completions of measurements, a Dimension Inspection Report (DIR) is prepared on your letter head giving PO No, PO date, specification No, name of the product and quantity. Dimensions are recorded in tabular form.

A specimen of DIR is placed at Annexure-6.

8.22: Hydraulic Pressure Test: It is also called Hydro-Test or 'Leak Test'

Many Product Standards specify Working Pressure and Test pressure, which is normally 150% of designated working pressure.

This test reveals invisible cracks in material, welded joints, Assembly of parts etc. Many a time Liquid Penetration test fails to detect hidden cracks, which develop when the component is stressed under working condition.

8.23: Equipment for Hydro-Test: (Based on author's experience)

(a) **Hand operated Pressure Test Pump** equipped with Shut-off Valve, Pressure Relief Valve, Pressure Gauges, High Pressure Hose and Water Tank. Capacity/rating of the pump should meet wide range of your requirements. You may require 3 or 4 pressure gauges with valid calibration certificates.

Thumb rule: Pressure gauges should not be used beyond 70% of their maximum capacities.

(b) **Job Holder:** Type and size of job holder depends on the size, shape and quantity of the products to be subjected to hydrostatic pressure. It also depends on test pressure, and whether it is internal or external.

i) Large products like Mud Suction Valves of Oil Field Equipment require two large circular plates lined with rubber discs and two or four long studs with washers and nuts. (Refer to Fig. 1 in Annexure-7). Smaller sizes of such fixtures are useful for medium sizes of products like Oil Throttle Valves

ii) Medium size products like Adjustable Orifices and Pipe unions can be tested in a specially designed vice. (Refer to Fig. 2 in Annexure-7)

iii) Thermowells: They are tested by applying external pressure and require Pressure Test Vessels.

For Medium Pressure thermowell, a vessel fabricated from seamless pipe and fitted with 5 or 7 numbers of Vertical Inserts is used. One such vessel is shown in Fig.3 of the Annexure-7.

For High Pressure Thermowells, vessel can be bored from a MS rod to accommodate only one thermowell, with top end fitted with an internally threaded flange.

For designing the pressure vessels, i.e. Schedule of Pipe and thickness of end plates, consult a pressure vessel designer.

NB: Pressure Test Vessels should be stress relieved after welding.

iv) Small-size products in large quantity: For testing small products like 3-piece unions and other pipe fittings in large numbers, it is economical to test them in Multi-Port fixtures

One such fixture is shown in Fig.4 of the Annexure-7 and its photo can be seen in Fig.6. This fixture was designed and machined by Mr. A. Rajaiah, senior most Turner of Value-Trek Engineers. This proves that age is not a bar for innovation. For 30 years, the products were tested one at a time using the Vice shown in Annexure-7. Now, 18 Nos. can be tested in one go!

v) In addition to the aforesaid fixtures, you may be required to find some other method of holding the job. It may be tested by clamping in the vice of power hacksaw or pressing the job in screw press. This method requires a specially designed plate at one end and a blind plate at the other end lined with rubber discs.

One such specially designed plate fixture is shown in Fig.5, Annexure-7

On completion of hydro test, a 'Pressure Test Certificate (PTC) is prepared, as per Annexure-8.

8.3: PERFORMANCE TEST:

Very rarely, some products require to be tested for performance. e.g. : Springs, Adjustable Orifices and Spray Valves for Dearators. These tests require Special Purpose Equipment, and performance is demonstrated by drawing characteristics like Load Vs Deflection for spring and Volume Vs Pressure Drop for Spray Valves.

8.4: SURFACE FINISH:

As a part of Final Inspection, the Inspector/QC Engineer checks whether the surface finish is as per specification. Normally such checks are 'visual' and sometimes by 'Thickness Gauge'.

After grinding, polishing and buffing, the external surfaces are coated by red oxide followed by one or two coats of enamel paint or the components are electro-plated or black-oxidized.

For the products which work in corrosive environment like the products of mud system of Oil Drilling Rigs, zinc silicate is used as rust inhibitor followed by 2 coats of 'Repack, High- build Polyurethane'.

In case of Cylinder and Piston assembly, the internal surface of cylinder requires honing

8.5: IDENTIFICATION MARKING:

The Inspector will verify whether the following data in marked on all products as per Specification or tags bearing required information are attached. Marking may be either by painting or by hard-punching.

- Suppliers Logo and ID No
- Name of product
- Nominal Size,
- Material grade
- Purchase Order No.
- Spec. No., Var. No.
- Component Code
- Inspector's stamp.

In case of complaint by the user, this information helps in identifying the supplier and claiming replacement during guarantee period or sending replacement.

A tag bearing P.O. No., name of the product in cartons, Item code and quantity is attached to box.

8.6: PACKING

The last part of Inspection (and QAP) is packing. Type and size of packing depends on:

(a) Shape, Size and Weight of the product.
(b) Destination
(c) Mode of transportation: By Road, Rail, Air or Sea.

The ultimate aim of packing is to prevent damages to products during transportation. The Inspector will check and be reasonably sure that the packing will protect the product(s) in transit.

Whatever may be the product, its size and mode of delivery, never send the finished goods in gunny bags, plastic sacks or pouches.

Here is the **list of packing materials and accessories**:

- Cartons for small and medium sizes of products.
- End Closures (Plastic Caps and Plugs).
- Cartons lined with moulded Thermocol for delicate items like electronic goods.
- Wooden crates for bulky items or for carrying cartons.
- Manila ropes for wrapping round the heavy products like Mud Guns, Suction Valves, etc.
- Paper and Paper shredding.
- Polythene sheets and silica gel.
- Hand Punched or Anodized Item Code Tags,
- Straps and Clips.
- Strapping/crimping tool

For Sea-worthy packing, your customer provides detail instructions.

Address Labels are pasted on the closed containers. Safety symbols are pasted or painted on four sides. For heavy containers, positions of slings/hooks are also marked.

8.61: Case Study-4: Planning to Avoid Delay in Delivery

In spite of meticulous planning, delivery of one consignment was delayed by almost one week.

In our Industrial Area, Monday is weekly holiday. On Thursday, the components were washed, dried and packed in cartons. They were then neatly arranged in wooden boxes lined with corrugated sheets. Consignee copies of dispatch documents were kept in the box. On Friday, I realized that the crimping tool borrowed by my friend was not returned; and he was on tour. He came back and returned the tool on Sunday evening. Monday was holiday.

Work was completed on Tuesday and consignment was kept ready for dispatch on Wednesday.

Can you guess the consequences of delay in dispatch?

In all documents, date was that of Friday whereas goods were being dispatched on next Wednesday. To set the record straight, the box was opened, documents were taken out and date was changed. Date was also corrected in other copies of documents. The box cover was nailed and box strapped once again. It has taken more than 6 man hours and lot of waste of packing material.

8.7: DOCUMENTS FOR INSPECTION:

Soon after entering your workshop for final inspection, the Inspector/QC Engineer will ask for the following documents as applicable to the product under inspection:-

   a) Mill Certificate for raw material(s)*

   b) Material Test Certificate (MTC).

   c) Dimension Inspection Report (DIR)

   d) Pressure Test Certificate (PTC)

   e) Heat Treatment Certificate (Time-Temperature graph)

   f) Radiography film (showing concentricity of bore with O.D.)

   g) Performance Test Report.

* MILL CERTIFICATE: Invariably, Inspecting Engineer on visit to your works will first ask for Mill Certificate for the raw material. Such certificates are given by the manufacturers to their main distributors who lift the entire lot (in tons) yielded by a single HEAT. Your requirement will be in 'Kgs. To come out of this vicious circle, you should make it clear in your offer that Mill Certificate is not available for small quantities and you shall provide Test certificate from a NAB Lab only.

Any remark by the Inspector on deviation and/or non-conformance should be referred to your customer giving reasons for non-compliance with request for acceptance. Your request is forwarded to Design/Engineering department. If the deviation does not affect the function, your request is likely to be accepted.

8.71: Case Study - 5: Study Tolerances Carefully

The drawing of a machined component gave 'FINE' tolerances on dimensions but what I achieved fell under 'NORMAL' open tolerances. I was confident that this deviation would not in any way affect the function or performance of the components. But I knew that this would not work unless the deviation was accepted by the customer. I followed the normal practice of writing to customer and got the deviation approved.

This approach to deviation may not work always.. You should either set right the defect or wait till the approval of deviation is received in writing.

9. DISPATCH

The main objective of dispatch is to ensure that the goods reach the destination safely and as soon as possible. This requires careful pre-planning concurrently with manufacturing activities.

9.1: PROCUREMENT OF PACKING MATERIALS: These materials of your requirement are not available off-the shelf and should be ordered well in advance. See the list of packing materials on page 41 and estimate the time required for their procurement.

9.2: PREPARING DISPATCH DOCUMENTS: It is a time consuming process. In fact, you can start preparing most of the following documents soon after receipt of purchase order.

here are two types of delivery: **1) Direct** to out-station destination and **2) Local delivery.**

You must generate an e-Way Bill valid for the date of dispatch

Before you move the goods out of your premises, you must ensure that five copies each of the following documents are ready in all respects:

1. Excise Invoice-cum- Delivery Note in Government approved format.
2. Delivery Challan. (DC)
3. Sales Tax Invoice
4. Inspection Report of QC Engineer or Third Party Inspector (TPI).
5. Dimension Inspection Report (DIR).
6. Pressure Test Certificate (PTC).
7. Guarantee certificate (GC). A specimen of GC is placed at Annexure-9.
8. Packing List.
9. Transit Insurance for out-station dispatches

In addition to above documents, you may be asked to provide:

10. Heat Treatment Record (i.e. Time-Temperature graph).
11. Performance Record.

For DIRECT dispatches, you are advised to put copies of all documents in a water-proof envelop, which, in turn, is kept in the packing box. One packing list is enclosed in water proof cover and nailed to packing box from outside.

You should list the documents enclosed in the box on Delivery Challan. If the space is not sufficient, attach the list to DC.

If the material is dispatched to your customer's customer, you will be directed to put a stamp on DC, Tax Invoice and Packing List: [MATL. SENT ON BEHALF OF (Yr customer's Name)]. [red colour]

Lastly, for Direct Dispatches, never forget to write Consignee's GST number on DC, Tax Invoice and Packing List.

10. BILLING

Having dispatched the goods, you would like to submit your Bill and get the payment as soon as possible. But money is the costliest commodity and will not come so easily.

The first requirement is the RECEIPT of GOODS (GR) from the recipient/Local Stores.

If the material was sent to your customer's Stores, the Inspector in "Inwards Goods' section will check randomly some (critical) dimensions. There is every possibility that he will detect some defects and reject the product(s). He will also check whether documents are in order. If some product is rejected, be prepared to go through yet another problem: Visit to Stores, knowing the nature of defect, replacement, inspection of newly manufactured product, and so on. Rejected item(s) is (are) sent to REJECTION STORES, and getting it (them) out for correction of small defect is another time consuming affair.

If the material was dispatched DIRECTLY, your bill will be taken for processing by the Accounts Dept. only after receipt of "CERTIFICATE of MATERIAL RECEIPT (CMR) from the consignee. This may take 4 to 6 weeks. (Refer to Para.(ii) on page 14.)

BILL should be prepared in four copies:

- a) First copy is marked: 'A/C Copy'. It should be accompanied by GR/CMR and submitted to A/C Dept. thru' PURCHASE.
- b) Second copy is marked 'PURCHASE'. Copies of documents (as required) listed on the previous page should be attached to this copy of BILL.
- c) Third Copy is 'OFFICE COPY' (O/C). Originals of all Dispatch Documents should be attached to O/C
- d) Fourth copy with dispatch documents is a SPARE set

Ideally, the process of preparing Tax Invoice (Bill) should start along with preparation of DC.

Entries in DC and Bill are common except the name of the document and Sl.Nos.

RECORDS: Referring to page15 (specimen of Order Book), you have data to fill up the columns on RHS, expect last 2 columns. While waiting for the Bill to be passed, you may complete the other entries in Order Book.

----

11. PAYMENT AND ORDER CLOSURE

On receipt of payment, you can fill the last two entries of Order Book and treat the PO as completed or closed.

TENDERING WITH INNOVATION

FOR MICRO, SMALL AND MEDIUM INDUSTRIES

PART-II: INNOVATION

12. INTRODUCTION

The process of tendering is governed by a set of strict rules, terms and conditions, leaving no scope for the bidders to change anything except price and delivery. If one has to work within such a rigid frame work, you may wonder where is the chance for applying innovative or creative ideas? It is a myth. Why and how? Read on.

Innovation means generation of novel idea that can be developed into a new product or process. For entrepreneurs, an innovation means an act of creativity to solve basic problems of reducing costs of Materials and Operations.

Rapid tides in technologies leave no viable alternative to developing and improving manufacturing methods and utilizing available resources to the fullest extends.

Imitators sprout overnight to rival an innovative process and product. In such a challenging scenario, innovation holds the key to leap-frog development.

PROBLEM: What is a problem and where to find it?

Anything that makes you <u>uneasy or disturbed</u> is a problem, and it is found everywhere man exists. For Entrepreneurs, 'cascading cost of operations and diminishing profit' is the basic problem.

(Discussion on 'Radical' or 'System' or Software innovations is excluded because they are beyond the reach of individual entrepreneur).

<u>**TECHNIQUES:**</u> How to solve a problem?

First requirement is SELECTION of problem:

1. <u>**Routine problems**</u> require routine solutions. (Feeling hungry? Eat food. Hinges are swishing? Put a drop of oil.)
2. <u>**Judgmental Problems**</u> have only one solution (I found a gold ring on the road. Shall I keep it? Shall we give Smart phone to kids? Solution to such problems is: YES or NO.)
3. <u>**Analytical Problem**</u> also has only one correct answer. (Find the missing number in: a) 51, 42, ..., 24, 15. b) $4^2 + 5^2 = ?$)

4. <u>Creative Problem</u> yields a large number of solutions:

[a) What are the uses of a Saree?

b) Divide two concentric squares into 4 parts of the same shape and equal area)).

This problem[1] has infinite solutions but 99.9% people stop after 3 or 4 familiar shapes because of their liking to tread the beaten path

c) Brass Tap is leaking. How many ways can you stop the leakage of water?

Ans.: Many ways and cost of one of them is nil. (See Book[2] by the author)

First, make sure that the problem does not belong to first three categories i.e. it is Creative type.

Second requirement demands removal of **<u>'MENTAL BLOCKS'</u>** or Inhibitors of innovative process:

[a) Why change it when it's working well?

b) Customer will not accept it, and so on[1].]

Third requirement says: Search for opportunities intentionally.

"Break out of established patterns in order to look at things in different way". Edward de Bono.

This quote is very useful in overcoming "Functional Fixation', viz.:

a) Drill machine is only for drilling.

b) Filament Lamp is only to give light.

(Self) Brainstorming:

Appropriate questioning technique will reveal multicity of solutions to a problem.

- What is it? What else can achieve this function?
- Why is it necessary? Can it be eliminated?
- Who should do it?
- Where should it be done?
- How should it be done?
- When should it be done?

Kipling said "I keep six servants who taught me what I know. They are:

What, Where, Why, Who, When and How."

It means subject every problem to searching questions. You must also use 'spur' questions like:

- Why not combine?
- Why not rotate?
- Why not reverse?
- Why not make it bigger?
- Why not make it smaller?
- Why not shift location
- Which different material can achieve this result?
- Is there something similar?
- What else and what else... and so on.

This is an endless list. You should prepare your own list of specific questions that helps in activating imagination in the right direction.

13. IMPORTANT NOTES

IMPORTANT NOTES:

1) Before coming to the "specifics", you are urged to read the following two books by the author which contain several 'Case Studies' on diverse techniques of innovation. In his 2nd book, very frequently reference is made to PROJECTS for details.

[1] **Techniques of Training in CREATIVE PROBLEM SOLVING.**   Authored by

[2] **TITBITS and BOUTS OF CREATIVITY.**   **R.G. Chaudhari**

2) After running his own SSI Unit* for 27 years, which he started after retirement from 33 years of service, the author sold his business to a third party under agreement. He is under bond not to disclose some technologies that are discussed in the 2nd book. However, some hints are given for solution.

(* Name of the Industry: 'VALUE-TREK' ENGINEERS (VTE), e-mail ID, etc. remain unchanged.)

CASE STUDIES

The author believes that **Specification of Product is not a sacrosanct** document. It can be changed.

Your customer will be happy to accept the change(s) that achieve the primary (and secondary) function(s) of the product or the changes that improve the quality and reliability, and at the same time, reduce the cost.

The changes may be related to:

- Material of Construction (MoC).
- Size and shape of parts.
- Elimination of parts.
- Design of the part.
- Manufacturing Method, and so on.

14. CHANGES IN MANY ASPECTS

You should look for opportunities to change MoC, Technology, etc. as shown in the Case Studies:

CASE STUDY-6: LIQUID LEVEL INDICATOR.

I have selected this example first because it involves all the changes listed above.

To understand the implications of changes, you should read fully the Project No.32 in the 2ⁿᵈ Book.

Also, you should not miss to study the Project No. 11* and **'Problem for You No.1'** in the said book, which are related to LL Indicators.

Liquid Level Indicator is mounted on OIL (or water) tank to estimate the quantity of liquid in the tank. Centre to Centre (C/C) distance ranges from 125 mm to 430 mm for one type of end connections, shown here, and 850 to 2000 mm for another variety (shown in Project -11*).

Referring to Figure on RHS, it shows final version of the LL Indicator of type-1.

The changes made in the specifications of original design were as follows:

1. MoC of end blocks (P.No.1), protective tube (2), hollow bolts (3), nuts (4) and washers (7) changed from Brass to stainless steel. (Saving = > 40% in cost of materials)

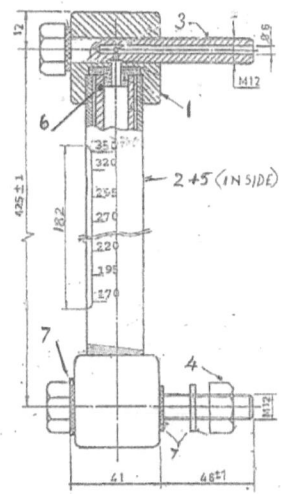

2. Size of end blocks reduced from 45³ to 40³, a saving of 30% in R.M.

3. Borosilicate gauge glass tube, (an imported item) was replaced with Acrylic (PMMA) tube after comparing their 'flexural' strength and 'rupture pressure'. It was found that Acrylic is superior to Borosilicate and its cost is about 30% of Borosilicate tube.

4. Indicator mounting bolts (3) were changed to 'Vectored Flow' bolts – **a new concept.**

5. Instead of machining these bolts from a hexagonal bar stock, Hex. Head was forged at one end of round bar, resulting in saving of about 60% in the cost of raw material and machining, vide Case Study -7.

6. Collared rubber bush (P.No.6) on both ends of Acrylic* tube ensure leak-proof joints.

This bush is a **new concept of sealing**, and was first designed by the author to overcome problem of leakage in 2000 mm long LL Indicator described in Project No.11, Book number-2. Thereafter, I made it a common feature in all types of LL Indicators designed by me.

7. Marking of graduation lines on the protective tube is normally done by engraving but it was done on lathe machine using a new fixture described under Project No, 37, Book No. 2, yielding a saving of about 20-25%.

NB: Common practice of sealing at both sides of end blocks using rubber and aluminum washers (P.No7) did not succeed 100%. I had to add some other part.

'I am bond by agreement not to disclose that part and its installation'.

I am happy to inform that LL Indicators designed by me form parts of VTE's customer's specifications.

CASE STUDY -7: 'SIGHT FLOW GLASS' INDICATORS.

While we are on the subject of Borosilicate glass, let us consider the Sight Flow Indicators.

These Indicators have two round windows on opposite sides and at right angles to flow of oil. The windows are fitted with transparent glass discs. The assembly is pressure-tested at 2 Kg/Cm² or 16 Kg/Cm², depending on line pressure.

The size of the discs varies from Ø 35 to Ø 63..

1) Change in Material specification.

The original Product Standards specified 'Borosilicate glass', which was an imported item; and the price was exorbitant. This made me think of alternatives. The alternatives were:

(a) Toughened Soda Lime Glass for high pressure and (b) Acrylic (PMMA) for low pressure applications.

The thickness of soda lime discs was determined by pressure testing the assemblies at 24 Kg/Cm². Several trials revealed that thickness of 6 mm was suitable for first 3 sizes and 8 mm for the higher sizes.

2 mm thick Acrylic disc withstood test pressure of 3 Kg/Cm² as against required test pressure of 2 Kg/Cm².

The change proposals to replace Borosilicate glass by Soda lime Glass (or Acrylic discs) were accepted by the customer.

This change in material specification of glass resulted in saving of approximately 70%.

2) Change in Process:

The Specifications say 'Body should be one piece'. Obviously, fabricated body is not accepted. Utilizing forging facility in the shop and machining the body, I found that the cost of this process is disproportionate to its functional value.

With experience in Investment castings, I made dies for wax patterns and got the bodies done by Investment casting process.

Saving in cost of material and machining was about 20 to 22%.

You will find most astonishing change in the material specification in the following CASE STUDY.

CASE STUDY NO.8: DT PIPE UNIONS- CHANGE OF MoC

Readers are advised to read 'Project No.26, Book No. 2' for details regarding genesis of project, method of investigation, etc. Highlights are as follows:

1. One of the several products that had come to me for indigenization was Dilo Type(DT) Pipe Unions, which were imported from Germany. [I was then In-charge of Prototype Development Centre (PDC) in BHEL].

2. After decoding the secret of leak-proof joints of metal to metal parts, I was directed by the Management to develop vendors. I furnished necessary sketches and instructions to Engineering and two parties.

After one or two years, I retired and started my own small industry to manufacture Import Substitutes developed by me in PDC, including DT Unions.

3. As per Product Standards of DT Unions, Carbon Steel (CS) and Stainless Steel (SS) unions mean all parts should be made of CS and SS, respectively. To avoid <u>seizing of SS parts</u> in service, they were coated with **Molybdenum disulphide**. This Technology was available at M/S TVS, Chennai but the cost of coating was almost equal to the cost of components.

There are about 30 verities of unions which are commonly used. Three of them are shown in the drawing.

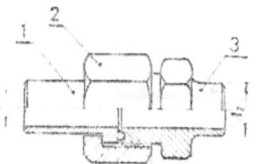
Figure 1: Straight union

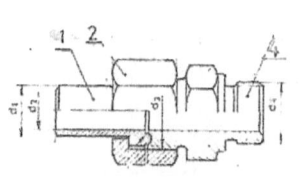
Figure 2: PMS Union

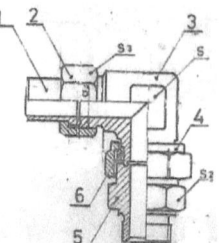
Fig3: Angle Swivel Union

1. Weld-on Stub. 2. Cl. Nut.   3. Weld-on Body.   4. PM Stud

In quest of alternative solution, I put the following questions to myself; and the answers turned out to be astonishing.

Questions:

1. If the threaded SS parts seize, being made of similar metal, why not use two parts of dissimilar metal?

2. Function of Clamping Nut (Part-2) is to clamp the two parts together (P.No.1 & 3 or 1 & 4). Then, why not make the nuts in Carbon Steel?

3. The body (P.No.4) is screwed into CS Casing. Why not make the body also in CS?

I made a few 'SS unions' with weld-on stubs in SS and other parts in Black-Oxidized CS; and kept them in water for about 3 months. The results were on expected line. Parts did not seize. I continued this exercise on 2 or 3 more unions.

Subsequently, on my suggestion, the specifications were amended by the customer to read as follows:

A) CS unions mean all parts are made of Carbon Steel.

B) SS unions mean only the stub (P.1) and weld-on body (P.3) should be made of SS and all other parts, in CS.

Advantages:

1. The costly process of coating Molybdenum disulphide is eliminated.

2. Costly Stainless material was replaced by Carbon Steel, resulting saving of about 45 % to 94% in Stainless Steel material based on the following example.

Exercise of this nature is called Value Engineering.

Example: To illustrate the percentage savings, I have taken three unions, shown on the previous page as reference. Their common feature is Nominal Pressure of 100 Kg/Cm².

	Fig.1	Fig.2	Fig.3
Percentage Ratio of SS to CS	55:45	16:84	6:94.

It may be noted that rate of SS321 is about five times the rate of CS:A105.

No wonder, the Head of Piping Engineering called this change as **revolutionary**.

CASE STUDY NO. 9: TOOL STEEL Vs MILD STEEL

(Please read full article in Book-2, Project No.33. Use of mild steel in place of tool steel is not covered there.)

Pressing 'Fingers' used in TG set are made from non-magnetic, (Manganese) steel flats. The cross sections (WxT) of the flats are 8x50, 12x50, 15x70, etc. Length of cut-pieces 'L' varies from 430 to 660. Raw material comes in nominal length of 5000.

While the flats were unloaded from a lorry, I observed that some flats were bent and some, twisted. Obviously, the material was soft like mild steel. They could be easily straightened under screw press.

For soft material, why Tool Steel is required for bending tools?

I made the bending tools by gas cutting them from 50 mm thick MS plate. Each set of Fingers require minimum two sets of Bending Tools: one for pre-bend of 2 mm and the other one for round bend.

In early 1990s, cost of tool steel was approximately three times the cost of mild steel.

Average weight of raw material required for one set of bending tool works out to 15.0 Kg. In due course, about 20 sets were made, each set yielding a saving of about 75% in cost of material and heat treatment.

Even after using the MS bending tools for nearly 25 years for bending thousands of pieces, no distortion or deformation was found in the tools.

15. SEIZE THE OPPORTUNITY: FORGING.

There goes the adage: "Innovate or perish".

One of the best methods to stay afloat in the stiff competitive market is to constantly look for opportunities to reduce **wastage of materials.**

Example: I was getting small parts forged from a nearby workshop. One morning, an uneducated but highly skilled Forge Smith working in the said works approached me with a proposal:

"The hammer on which your jobs are forged is built by me. Give me money to fabricate similar hammer, and space to install it. I will do your jobs on priority".

In due course, Value-Trek had its own forging facilities with 250Kg. hammer; and saved tens of thousands of Rupees by reducing scrap and machining time

Before dwelling in further case studies, let me tell you that our Forge Smith has developed two- part "Jumping dies" to reduce the diameter of a round rod progressively.

For example, if a rod of Ø 72 mm is to be reduced to Ø 20 mm, first Ø 72 is cut in to pieces and then each piece is reduced to ,say, Ø50 mm; next Ø50 is reduced to Ø30 and finally, Ø30 to Ø 20.

The finished rod is so smooth as if it was 'drawn'—no parting lines and no ovality.

----

16. WASTE REDUCTION BY FORGING

Normally, specifications say: 'Machine the product from the bar stock'. Nowhere it is mentioned that the product should not be machined from the forging. Therefore, the product is commonly machined from the bar stock (without forging). This results in generation of heavy scrap (wastage of material) and loss of machine time.

The following examples show the way of reducing consumption of raw material, cutting tools and machining time, thereby reducing the cost of product.

16.1: CASE STUDY-10 SQUARE BOLT (Carbon Steel):

The required size raw material is SQUARE 98 but the nearest size of raw material available in the market is 102x102.

There are two methods of making the bolt shown here.

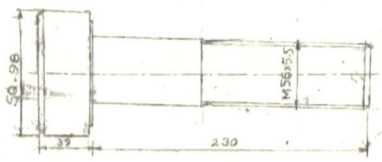

1) Machine the bolt from a square bar stock.

2) Forge the round part from the square part.

In both cases, square head is common. Hence, it is not considered in the following calculations.

- Weight of Square bar to machine the round part of Ø56x230 is 19.2 Kg.

- To forge Ø56x230, you need Square Bar of length 55.5 mm, weighing 4.5 Kg.

Net Saving = 14.7 Kg. Or 76 %

16.2: CASE STUDY-11: HEX.HEAD BOLT (Alloy Steel):

The required hexagonal head bolt is shown in the drawing on RHS.

Calculations are similar to above example.

- Weight of Hex. Bar to machine the round part of Ø30x160 is 2.30 Kg.- **To forge Ø30x160, you need Hexagonal Bar of length 55.5 mm, weighing 0.98 Kg.**

Net Saving 1.27 Kg or 57 %

16.3: CASE STUDY-12: VECTORED FLOW BOLT OF LEVEL INDICATORS: (SS304)

(Refer to drawing of Case Study-1. Item No.3)

In this example, instead of forging round from hexagonal bar, Hex. Head is formed by forging at one end of round bar

(a) If the Bolt is machined from hexagonal (A/F 19) bar:

Weight of Hex. Bar = 0.282 Kg..........(a)

(b) If Hex. Head (A/F 19 x 9) is formed by forging , it requires 25 mm of round Ø12.

Therefore, total length required= 25+88 = 115 mm.

Weight of round bar = 0.136 Kg.........(b)

<u>Saving in Raw Material = (a) – (b) = 0.146 Kg. or 51%</u>

This appears to be a drop in ocean. Please, remember: Drop by drop, ocean is formed.

The order quantity in **one case** was 130 Nos.(260 bolts). The cost of machining the bolt from Hex- bar was much more than the saving in material.

16.4: CASE STUDY-13: 13CrMo44

From my experience, I can cite tens of cases wherein not only the wastage was reduced but also the problem of scarcity of some special material was overcome by forging techniques. One such example is given below.

This material (13CrMo44) of German specification is used for high temperature service.

This grade is specified in customer's Product Standard for DT unions. Required sizes of bar stock were: Rounds Ø22, Ø30 and Ø36, and Hex A/F 24, 27, 32, 36, 42 and 48, all in very small quantities.

My long search through suppliers in Mumbai and Chennai and scanning Yellow Pages proved futile. Finally I located one dealer in Mumbai who had a round bar of Ø 76 x 1.2 M.

The rod was cut in to pieces and the pieces were forged progressively to required diameters. Round bar was then cut in to pieces and forged in Hexagonal dies of required sizes.

On successful completion of the order, DT unions of Alloy Steel were declared as 'Indigenized' by our customer. For the last nearly 25 years, 'Value-Trek' is the only supplier of these unions, saving precious foreign exchange and earning handsome profit.

16.5: CASE STUDY-14: THERMO WELLS: (Refer to Annexure-2, Exmple-2, for shape and sizes of few thermo wells. Also, read Project No. 27 in the 2nd Book[2])

There are many verities of thermo wells. Commonly used 'Wells' are machined from SS316 rod of Ø 45. The table given below shows the savings in percentage:

S.No.	Bore Ø	Stem OD x length	If stem machined from		Saving	
	mm	mm x mm	Rod Ø45	Forged stem	Kg	%
01	6.5	12.5 x 150	1.97 Kg.	0.94 Kg.	1.03	52.3
02	6.5	12.5 x 250	3.22 Kg.	1.07 Kg.	2.15	66.8
03	8.5	15 x 150	1.97 Kg.	1.02 Kg.	0.95	48.2
04	8.5	15 x 250	3.22 Kg.	1.22 Kg.	2.00	62.1
05	12.5	19 x 150	1.97 Kg.	1.16 Kg.	0.81	41.1
06	12.5	19 x 250	3.22 Kg.	1.46 Kg.	1.76	54.6

Let us go back to Book[2], Project 27. Soon after starting production in my Unit in July 1988, Manager, Vendor Development (BHEL) urged me to enter in to business of Thermo wells. Within a year or two, production stabilized and I won almost all tenders for thermo wells. In spite of being L1, my customer asked me to offer at least 5% discount. I requested one week's time to think over the matter.

Here is how I looked for opportunity to innovate and earn handsome profit for many years.

I was, like my competitors, machining thermo wells from a rod Ø 45 and getting the rod bored on Gun Drilling machine. (And still my quotations were L1.)

After calculating scrap generated on 10-12 pieces, I machined a few thermo wells by forging stems (Saving = 40 to 45% of raw material). Also, I developed a low cost method of drilling bores, accruing a saving of about 30%.

I offered 25% discount! My customer was surprised, and asked me to check and re-check.

I made the offer on condition that the rates should be treated as valid for two years, with provision to extend it for one more year. *It was accepted.* Thereafter, I supplied hundreds of thermo wells.

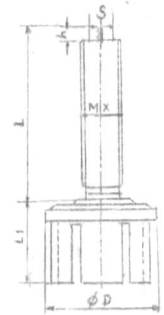

16.6: CASE STUDY-15: THROTTLE SPINDLE: I used the forging technique in the manufacture of Throttle spindles of Oil Throttle Valves and saved approximately 50 % of the cost of raw material and heat treatment charges.

In the below table, threaded stem only is considered.

S.NO.	Size	Threads	Raw Material		Forged stem		Saving in	
			Ø x L	Wt.-Kg	Ø' x L'	Wt.-Kg	Wt.(Kg)	%
1	**ND20**	M16	40x94	0.93	18x96	0.19	0.74	79.5
2	ND50	M30	65x144	3.75	32x146	0.92	2.83	75.5
3	ND80	M36	95x182	10.13	38x185	1.65	8.48	83.7

Savings in the last column of Table accrued due to forging of threaded screws.

If you consider the whole spindle, **real** savings are much more than that shown in the Table.

Referring to ND80, dia. D is 95 mm but the nearest size of round available in the market is 100+ and total length required is (L1+L+ 5) =73+182+5 = 260 mm. Weight of this piece =16.5 Kg. If the raw material Ø 100 is forged to rough shape {(Ø95x73)+(Ø38x185)}, its weight works out to 4.06+1.65 = 5.71,say, 6 Kg. Net saving=16.5-6.0 = 10.5 Kg (63.6%).

----

17. CHANGES IN TECHNOLOGIES

Economics of operations in MSME demand minimum investment in machinery and equipment. To meet this demand, entrepreneurs should think of utilizing available resources for operations other than that for which they are designed. The concept that Lathe is for turning only; Milling Machine is for milling only; Pillar Drill is for drilling only and so on, hinders generation of novel ideas.

17.1: CASE STUDY-16: MILLING OF SQUARE HEAD

Before taking up alternative uses of machines, let us see how a large square (like the one in Case Study-4) is machined on Milling Machine.

a) You may use an Indexing fixture (which is very heavy and needs time to set up and operate).

b) You may use a Try Square or a Set Square to ensure that two adjoining flats are at right angles. It relies on operator's visual judgment.

c) For small jobs like Square Head machine screws, a square-body chuck is used. Here, the threaded part remains outside and almost equal length goes inside the chuck. After milling the square head close to chuck, the inside part of the job is cut off., entailing the loss of nearly 40% of material.

While observing the way the machinist was setting the job using Try Square for milling square head of a big bolt, a spark of idea ignited; from which the following technique emerged.

Novel Solution: Run an accurately milled QUARE NUT on threaded spindle up to, say, half of its length and lock it with Hex-Nut. Across corner distance of lock nut should be less than A/F distance of the square nut. Hold the square nut of the assembly in milling vice and mill the top side. Turn the assembly through 90° and mill the second side; and continue till all four sides are milled.

This method does not depend on skill of operator; and saves considerable time of milling machine.

17.2: CASE STUDY-17: MILLING SQUARE HEADS on DRILLING MACHINE:

There are various categories and types of industrial valves. Here our discussion will be limited to valves manufactured to order. Such items are: Oil Throttle Valves, Adjustable Orifices, etc.

How to go about milling square heads?

Buy a cross slide of a lathe with dove-tail base and fix it on the work table of Pillar Drill.

Next, bore a taper sleeve to accept the straight shank of Square End Mill. Fix the end mill in sleeve with a screw (or braze it).

Run a square nut on the threaded stem and lock it with Hex. Nut, as explained in Case Study -9.

Fix the job in the vice of cross slide and mill the square head at the end of threaded screw.

This saves considerable time of milling machine and consequently, reduces the cost of machining square.

The cross slide can be put to use for other operations such as milling keyways, cutting slots, etc.

OTHER USES OF ABOVE TOOL:

I was in the business of supplying pipe fittings, including shell type pipe clamps, In case of pipe clamps, two click nuts are press-fitted in counter-sunk holes in base plate.. I got the two counter-sunk holes done using End Mill on Drilling M/C. Quantities was in hundreds.

Holes in flanges can be spot-faced' using the above tool on drill machine.

(I made about 10 tools with end mills Ø 6.5, 8.5, 10.5, etc, up to 24.5 mm).

17.3: CASE STUDY-18: USE OF END MILLS ON LATHE MACHINE.

End mills fixed in taper sleeves can be used in several ways on a lathe machine:

1. I used to supply SS 90° Elbows in hundreds. After drilling forgings from both sides, drill cone was left at least on one side. This was not acceptable to our customer.

I solved the problem by replacing square end mill by keyway end mill fixed in the taper sleeve and completed the jobs on lathe, thereby saving hundreds of hours in setting jobs on milling machine.

2. When blind holes are drilled in round jobs, drill cone are formed at the ends of bores. In some cases like thermo wells, it is specified that the ends should be flat. Usual practice is to grind the end of drill flat; and remove the drill cone. After completing flat drilling, the original cone shape is restored by grinding. In this process the life of drill is reduced by nearly 40%. Instead, if square end mill fixed in taper sleeve is used, wastage of drill and time for its grinding and re-grinding are eliminated.

17.4 CASE STUDY NO. 19: DT STUBS (USE OF END MILL ON LATHE).

The drawing shown on RHS is called WELD-ON STUB. It is a part of every type and size of DT unions; and the requirement of each size runs in to hundreds.

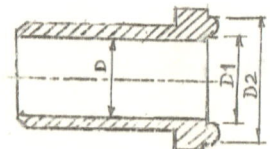

Size of diameter D1 is proportional to the bore D. Size of ØD varies from 3.5 to 38 mm.

Before machining R1.5 profile, diameters D1 and D2 must be turned within ±0.01. Turning D2 was not a problem but D1 was taking very long time.

Solution: I got commonly used sizes of End Mills ground to the required diameters, fixed them in taper sleeves and milled ØD1, 1.5 deep.

(With availability of CNC machining at reasonable rates, the use of End Mills is now restricted to small batch sizes).

17.5 CASE STUDY-20 CORNER CUTTING OF FLATS AT 45°. [Refer to Project No.33 in Book[2]. A part of this article (viz. Problem-3) is produced below].

The flats are made from Manganese Steel and are called 'FINGERS'. They can be cut only by abrasive cut-off wheel (or Carbide tipped slitting saw). Their cross sections are: 8x50, 12x50 15x70, etc., and lengths vary from 430 to 660 mm.

One of the operations was to machine one corner to R10.

The Route Card says: Hold 4 or 5 flats in machine vice and gang mill the corners at 45° up to 6 mm, and then grind the corner of each flat to R10.

Carbide tips of T-Max cutter were wearing out fast. Operations on milling machine are always costlier than on any other machine.

Alternative use of Cut-Off Machine: This machine is exclusively used for straight cutting of pipes, rods, angles, etc. When the cut- off wheal wears out to about Ø200, it is scrapped.

I found an alternative to milling corners of flats using cut-off machine and discarded wheels.

<u>Method:</u> Rotate the Machine head through 45°. Clamp the job in flat position. Use a discarded wheel and cut off the corner (6x6). Then grind the corner to R10.

It takes about 10 seconds to cut the corner; and cost of consumable tool is zero.

(If a wheel of bigger diameter is used, it bends and gives taper cut. Also the danger of wheel breaking exits

17.6 CASE STUDY-21: ADJUSTABLE ORIFICES.

To my knowledge, Adjustable Orifices with **Square Body** are in use for over 55 years. (See Fig.1 & 2). Commonly used material of construction for Body and throttle screw is SS 304.

Finish sizes of bodies are as follows:

<u>Cross Section x Length</u>

1. 50 x 50 x 60

2. 60 x 60 x 80

3. 80 x 80 x 110.

Problem:

Cross section of square bars available in the market is not perfect square. One of the methods is to buy the bar of nominal size, cut it in to pieces and forge them in the die.

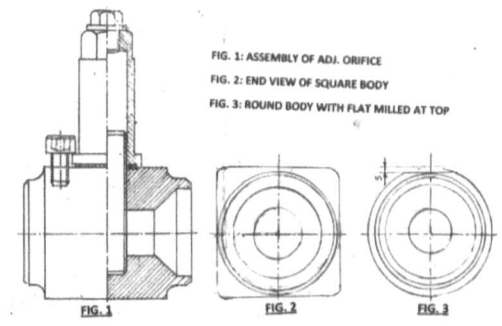

FIG. 1: ASSEMBLY OF ADJ. ORIFICE
FIG. 2: END VIEW OF SQUARE BODY
FIG. 3: ROUND BODY WITH FLAT MILLED AT TOP

FIG. 1   FIG. 2   FIG. 3

After stress relieving, six sides are milled to the required size.

<u>Solution:</u> While observing this tedious process, I was wondering why somebody had not thought of redesigning the product during past 55 years. By looking at the end view (Fig.2), idea of 'Round Body' with flat milled on the top (Fig. 3) stuck my mind. Round body saves more than 20% of material and eliminates lengthy process of forging, H.T. and milling.

Saving: Overall saving of about 40% accrues.

CONCLUSION:

The Case Studies given above may not be related to your product(s) but I am sure that the core message is conveyed effectively, viz.:

- Specifications can be changed, partly or wholly.
- Intentionally, search for opportunities and grab them.
- Innovation in tendering is the key to astonishing success.

TENDERING WITH INNOVATION

FOR MICRO, SMALL AND MEDIUM INDUSTRIES

ANNEXURES
(SPECIMENS)

SPECIMEN: SUMMERY CHART OF DIRECT AND INDIECT EXPENSES

Code	Expense heading	Code	Expense heading
	DIRECT EXPENSES		**DEPRECIABLES**
1	Raw Materials	30	Plant and Machinery
2	Purchases for Trading	31	Patterns, Dies and Moulds
3	Consumable Store	32	Q.C. Instruments, Test Equipment
4	Packing Materials	33	Electrical and Electronics
5	Transport and Freight	34	Office Furniture and Equipment
6	Wages, Incentive, Bonus. Rewards	35	Vehicles
7	Electricity and Water	36	A/C and Air Coolers, etc.
8	Job Work paid		
	INDIRECT EXPENSES		**MISCELANNIOUS**
9	Remuneration to CEO		You may note the following Codes against the following heads in your Cash Book.
10	Salaries (*Indirect Employees*)		Deposits,
11	Rent	37	Advances
12	Employee Welfare (EW)	38	TDS remitted to Department, etc.
13	Fuel/Conveyance, Tour & Travel	39	
14	Stationery		1. As a part of Annual Planning, you should maintain your own Cash/Day Book with RHS page for DEBIT and LHS for CREDIT, each page with 5 columns.
15	General Expenses		1st Col. = Date of transaction
16	Communication (Phone, Post, etc.)		2nd Col. = Code from this chart.
17	Repairs and Maintenance		3rd col. = Description of transaction
18	Rates and Taxes		4th col. = Cash received/Paid
19	Interest to Bank		5th Col. = Amt. received/Paid by Cheque or Online transaction
20	Bank Charges (Services, Chq. Book)		
21	Inspection Fee		2. Monthly summary sheet(s):
22	Testing and Development		Take 1 or 2 A3 size ruled sheet(s). Divide 1st sheet in to 31 columns:
23	Business promotion (Ads. Gifts, etc)		1st col. = Sl. No.
24	Vehicle Upkeep		2nd Col. = Month
25	Insurance		3rd to 31st Cols. = Code + expense
26	Donations		
27	Delayed Delivery Penalty (DDP)		
28	Tax Deducted at Source (TDS)		
29	Untitled Expenses		
Note:	In consultation with your Tax Advisor, you may add, split or combines some		

EXAMPLES OF COSTING SHEETS

Example -1: Multi-part product

Product: Liquid Level Indicator, C/C 127, Spec. abcd--- , SS 304, Qty. 126 Nos.

Enquiry No. KMN-9999 Dt: 10-12- 201d. Qtn. No. 4567 Dt: 01- 09- 201d.

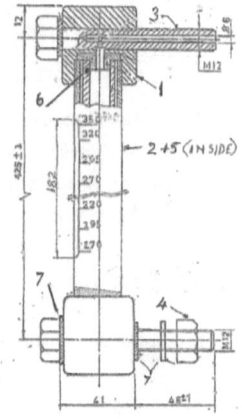

A. Raw Materials:	Reqd. (Nos.)	Qty.	Rate (Rs)	Cost (Rs.)
1. End Block: SS, Sq.42 x 45	2	1.3 Kg.	235/Kg	305
2. SS Tube, ERW, OD25x2Thk *	1	0.25 "	340/Kg	85
3. Vectored Flow Bolt, M12x90 Lg 2	0.23 Kg.	190/Kg	44	
4. HEX. Nut M12 (Bought out)	2	2 Nos.	10 each	20
5. Acrylic Tube OD 20x2Thk	1 500 mm#	110/Mtr.	55	
6. Sealing Bush, Neoprene(B/O) 2	2 Nos.	7.50 each	15	
7. Al, SS and Rubber Washers	4 each - - - - - - - - - - - - - Lump sum			24
8. Polythene spacer* & plug*	2 each - - - - - - - - - - - - - Lump sum			12
			Total 'A'	560

*(Not shown in Figure). # Min. purchase Qty.

B.Operation Cost: (Based on Time x Rate in the Master Register of Operations)

	Hacksaw		Abrasive Cut- Off m/c		Cost
1. Cutting:	Sq.42x45	Hex.19 Bar	SS Tube OD 25	Acrylic Tube	(Rs.)
Rate:	12 x 2 Nos.	1.5 x 2	1.5 x 1	1.5 x 1 -- -- --	30

2. Machining: Facing (6 sides), Drilling and threading of Item-1 -- -- -- -- -- -- — 59

3. Facing and threading of Item- 2 and 3 -- -- -- -- -- -- -- -- -- -- -- -- -- -- -- — 28

4. Engraving of graduation lines and numbers. -- -- -- -- -- -- -- -- -- -- -- -- -- — 43

5. Slot Cutting in SS Tube –Item-2. -- -- -- -- -- -- -- -- -- -- -- -- -- -- -- -- -- — 15

6. Cut relief and 3 holes in Item-3 -- -- -- -- -- -- -- -- -- -- -- -- ---- -- -- -- -- — - 15

7. Rounding of edges of end blocks Item-1 by grinding and

    sanding: 18x2 Nos. -- -- -- -- -- -- -- -- -- -- -- -- ---- -- -- -- -- -- -- -- -- -- — 36

8. Deburring, polishing and buffing of SS Parts. -- -- -- -- -- -- -- -- -- -- -- -- — 24

9. Detergent Wash-- -- -- -- -- -- -- -- -- -- -- -- -- ---- -- -- -- -- -- -- -- -- -- -- — 2

10. Assembly and Pressure Testing-- -- -- -- -- -- -- -- -- -- -- -- ---- -- -- -- -- — 33

11. Packing- Carton -- -- -- -- -- -- -- -- -- -- -- -- -- ---- -- -- -- -- -- -- -- -- -- — - 5

Total 'B' 290

Manufacturing Cost A + B = 850

Overheads @ 20% = 170

Factory cost = 1020

Add profit margin @ 40% = 408

Selling Price Rs. 1428

Round off and quote Rs. 1430/-

Expected Order Value for 126 numbers is Rs. 1,80,180/-

EXAMPLE -2: SINGLE-PART PRODUCT

Product: Medium Pressure Thermowell, Spec. MP 12345, MoC : SS 316, Order Qty. 120 Nos.

Enquiry No. KMN-9999 Dt: 15-12- 201d, Quot. No. 5678 Dt: 01- 09- 201d. Due Date: 05- 09- 201d.

Description: Thermowell, Screwed, M33 x2 (M), M20x1.5, Head Dia. 45, and A/F 36 (Constant Dimensions).

(Variable Dimensions are: Bore Ø C = 6.6, 8.5, 12.5 and 14.5 and Outside Ø D = Ø C + 6.5. Under each bore dia. there will be, say 4 Immersion lengths L1 = 150, 175, 250 and 325. For each immersion length, there may be 3 extensions lengths N1, viz. 27, 75 and 100. In other words, for each bore size there will be 12 variants.)

In description, dimensions against any one variant will given.

As an example, 12 variants of bore dia. 8.5 are shown in the following table and any one of them will figure in the description:

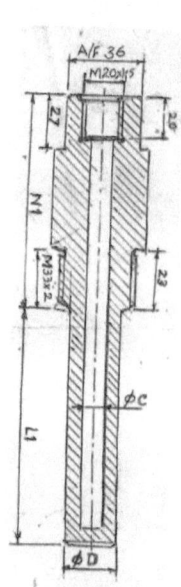

Var. No.	Bore Ø C	Stem Ø D	L1	N1	Item Code (Assumed)
01	8,5	15	150	27	MP881-234
02				45	-239
03				100	-244
04			175	27	-249
05				45	-254
06				100	-259
07			250	27	-274
08				45	-279
09				100	-284
10			325	27	-289
11				45	-294
12				100	-299

Many a time, internal threads M20 are changed to ½"NPT or ¾" NPT.

Thus, for Medium Pressure Thermowells (40 Kg/cm²), there will be about 75 variants.

Similarly, for High Pressure Thermowells (250 Kg/cm²), machined from the same Ø 45 bar stock, there will be about 75 variants.

Idea behind giving this information is to reduce time for calculating cost of operations.

First, list out common operations and their costs as follows:

A. CONSTANT COST: (Valid for 2014-15).

Sl.No.	Operation	Cost (Rs.)
01	Cutting Ø45 SS Rod, Deburring & Centre drilling (16 + 2 + 3)	21
02	Machining of Threads, Maintaining total length	30
03	Flat drilling of bore end, Cut off 'centre' and facing of ends	22
04	Flat milling A/F 36	20
05	Dimn. Inspection, Data Punching, Pressure Testing (5 + 5 + 24)	34
	Total	127

B. VARIABLE COSTS: (Valid for 2014-15).

Cost of raw material, External Turning, Pre-drilling, Deep drilling, Polishing, buffing & washing (Average of Max. & Min. lengths), etc. constitute variable cost of manufacturing Thermowells.

C. SELLING PRICE: Sum of A and B gives manufacturing cost. Add over-head Expenses, profit margin and contingency to get the selling price.

Note: The costliest variable operation is 'deep- drilling. You can further reduce the time for Calculating time for deep drilling by establishing standard times for drilling holes in steps, say, up to 80 mm, 81 to 150, 151 to 240 and so on.

SPECIMEN OF COSTING TABLE OF HP THERMOWELLS

Sl. No.	Item Code	L= L1+N1	Raw Material @Rs.270/Hg		Const. cost	Extrnl. Turning	Bore	Polish Per mm	Oprn. Cost	Mfg. Cost	Overhead (20%)	Factory Cost
		R/O	Wt.	Cost		0.20/mm		0.04	(Rs.)	(Rs.)		
1.	ABC	175	2.18	589	127	34	35	7	203	792	159	951
2.	ABD	225	2.81	678	127	45	75	9	256	934	187	1121
3.	ABE	250	3.12	842	127	50	90	10	277	1119	224	1343
4.	ABF	275	3.44	929	127	55	105	11	298	1227	245	1472
5.	ABG	325	4.06	1096	127	65	135	13	340	1436	287	1723
6.	ABH	350	4.38	1183	127	70	140	14	351	1534	307	1874
7.	ABJ	-----	----		127	-----						

 (in RED ink)

Up to 100 mm @ 0.20/mm

101 to 200 mm @ 0.40/mm

> 200 mm @ 0.60/mm

There are many types/varieties of thermowells such as:

- Screwed, Stainless steel, Medium Pressure, Round Head and Hex. Head;

- Screwed, Stainless steel, High Pressure, Round Head and Hex. Head;

- Screwed, Stainless steel, Non-steam;

- Flanged;

- Thermowells come with various types of Instrument connections.

- MoC may be Stainless Steel or Brass.

It is advisable to prepare a list of "Constant Costs" and Charts of "Variable costs" of operations for thermowells covered by each specification, as shown in the specimens. This one-time exercise saves considerable time in preparation of price bids.

ANNEXURE-3

SPECIMEN: ORDER ACKNOWLEDGEMENT

(To be typed on your letterhead and sent as an attachment to e-Mail)

To
Your Customer's (Co.'s)
Name and address.

Dear Sir/Madam,

> Attention: Mr./Ms ....,
> (Name and Designation of Purchase Executive).
>
> Sub: Order **Acknowledgement.**
> Ref: P.O. No. ABC999999 Dated DD/MM/YYYY.

We (I) acknowledge with thanks the receipt of the purchase Order cited under reference, and received on DD/MM/YYYY .

We (I) have noted, and recorded the Delivery Date as (date/month/year) in our (my) Order Book, and we (I) will try our (my) best to deliver full quantity (of each item) of ordered materials well before the due date.

Thanks and regards

Name and Designation
(Authorized Signatory)
Mobile Number.

You may like to get the specimen printed on A5/A4 paper, leaving spaces for variables for filling by hand

SPECIMEN: LIST OF TOOLS

1. HAND TOOLS

In order to enhance productivity of human efforts, you might have procured most of the following hand tools of required types and sizes:

Vices, Spanners, Wrenches, Hammers, Pliers, Files, Screw Drivers, Steel tape and Scales, Hacksaw, Cutting Tools, clamps, Instruments, : Allen Keys, Try Square (2-3 Sizes), Scraper, Pincer, Grease gun, Centre punch, Bearing Puller, Spirit level, Set Squares, Crimping tool, etc.

'Bradma' Roll Marking Machine with interchangeable steel types, holder for 14 steel types and letter & number punches.

A platform trolley either shop-made or bought-out and a Screw Press.

Drilling Jigs, Fixtures and similar devices help in reducing operation times and enhancing quality of products.

2. PORTABLE POWER TOOLS:

The common types of portable **power** tools are powered by (1) Electric motor and (2) compressed air (Pneumatic).

In Engineering Workshops, commonly used power tools are:

- Electric power drill with attachments for drilling, de-burring, polishing, buffing, etc.
- Straight Grinder with 100 mm grinding stone (wheel).
- Angle Grinder with grinding wheel and cut-off wheel
- Cut-off machine with slitting saw and abrasive wheels.
- Pneumatic die grinder.
- *Flex shaft grinder* with mounted abrasive points.
- Belt-cum- disc sander.
- Vibrator Engraving Tool: Very useful for engraving Item Codes on Al tags in emergency.

3. MACHINE TOOLS:

Your factory must have been equipped with most of the machines listed below, and you might have selected their types, sizes, controls, etc. based on your actual or perceived requirements.

1. Lathe Machine(s).
2. Milling Machine(s).
3. Drilling Machine(s)
4. Hacksaw Machine.
5. Abrasive Cut-off machine.
6. Polishing and buffing machine.
7. Air Compressor
8. Grinder (s)
9. Motorized De-burring and Polishing Drum.
10. DC Welding Machine.
11. Searing Machine.
12. Folding Machine
13. Forging Hammer
14. Hydraulic Press
15. CNC Lathe

QAP SPECIMEN (WITH A FEW SAMPLE ENTRIES)

QAP SPECIMEN (with a few sample entries) MANUFACTURING QUALITY PLAN ANNEXURE-5

VVendors Name & Address					
	Customer's Name	P.O.No.		QP No. :	Date:
	Project:	P.O.Date:		Rev.No :	Date
	Product:	Spec.	Rev.		Page 1 of ..

Component	Characteristics	Class	Type of check	Quantum Of check	Reference Document	Acceptance Norms	Format of Record	AGENCY D	P	W	V	Remarks
Raw Materials and Bought –out Items												
1. Body	Chem. & Mech. Properties	Critical	Chem. & Mech Properties	One/Lot	ASTM –––	ASTM A182 F321	Lab. TC		✓	2	'– –	Review by TPI
2. Cap Nut	–Do –	Major	"	. "	"	SA105	Lab. TC		✓		2,1	"
3.												
In-preocess Inspection												
1. turning	Dimension conformance	Major	Measurement	10 % random by TPI*	Specification	Spec.	IIR		✓			*100% by Vendor
2. Flat Milling	"		"	"	"	"	IIR		✓			Review by TPI
3. Threading	"		"	"								
4.												
Final Inspection and Tersting												
1.Dimension Measurement	Dimension inclu-ding Threads	Major	Measure	10 % andom by TPI*	Specification	Spec.	IIR		✓			*100% by Vendor
2. Hydro–Test	Leak Test	Critical	Measure									

Surface preparation, Painting, Marking, Preservation and Packing as applicable will be additional rows

Notes: **Format of QAP provided by your customer gives LEGEND (meaning of D,P,W,V).There will be spaces (hereunder) for signatures with stamps for Vendor, QA Engineer and Customer.**

ANNEXURE-6

SPECIMEN : DIMENSION INSPECTION REPORT (DIR)

No.(Item Id)   Date: dd/mm/yyyy

P.O. No. ABCD78910   PO Date: --/--/----

Specification: (Enter Spec. No/Rev. No/Var. No.)

Product: Flow Regulating Valve: ND 40, ND60 & ND80

Quantity (NOs.) =   8   3   2

R=Required   A=Achieved

....

PO S.No. 1

Body (Casting):	a1	a2	D2	D3	D4	D5	b
R	20.9	26.7	32	32	38	47	130± 1
A	20.9+0.2	26.7±0	32+0.4	32+0.05	38±0.1	47±0.2	130± 1

2. Throttle Spindle	b2	d7	d4	d1	L
R	2.5	3	26	31.9	28
A	2.5±0.05	3 ±0.0	26±0.1	31.9±0.1	28±01

3. Cover	d1	d3	d4	h1	h2
R	38	35	80	12.5	10
	38-0.1	35±0.05	80+0.5	12.5±0.1	10-0.1

....

Repeat the above Table for the remaining two variants.

....

Signature and stamp   Signature and stamp
of Inspecting Agency   of Supplier/Vendor

SPECIMEN: HYDRO-TEST DEVICES

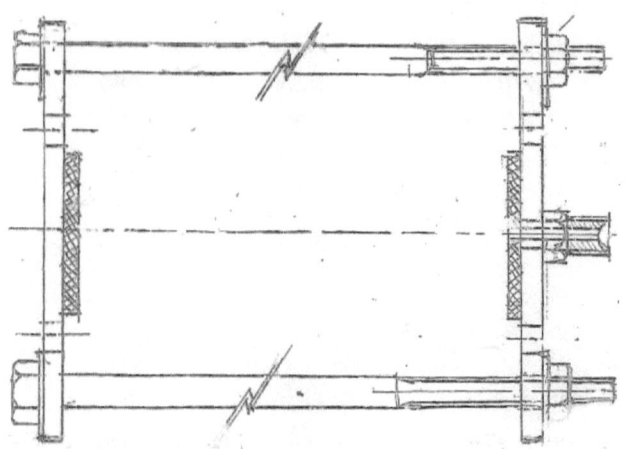

FIG.1: 2-PLATE DEVICE

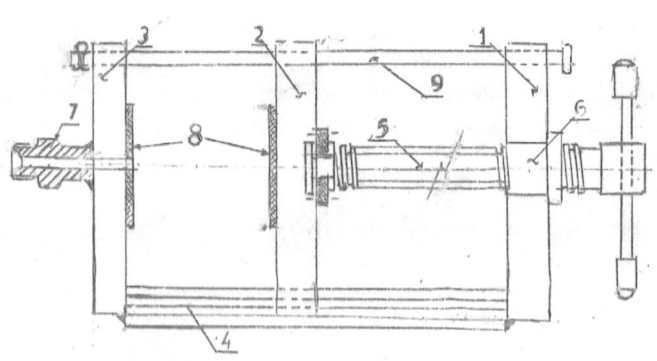

Fig. 2 PRESSURE TEST VICE

1 & 2 = END PLATES (26 x 100 x 175. 3. MOVING PLATE. 4 BASEPLATE WITH V-GROOVE 5. LEAD SCREW. 6. BOX NUT. 8. RUBBER DISCS (4 THK.) 9. GUIDE ROD Ø 12.

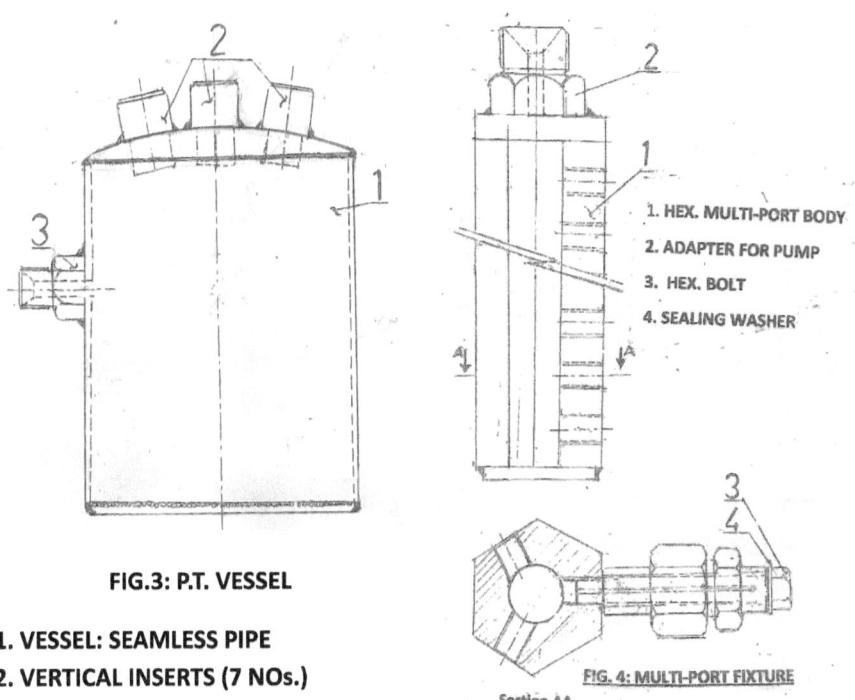

FIG.3: P.T. VESSEL

1. VESSEL: SEAMLESS PIPE
2. VERTICAL INSERTS (7 NOs.)
3. ADAPTOR: PUMP CONNECTION

1. HEX. MULTI-PORT BODY
2. ADAPTER FOR PUMP
3. HEX. BOLT
4. SEALING WASHER

FIG. 4: MULTI-PORT FIXTURE
Section-AA

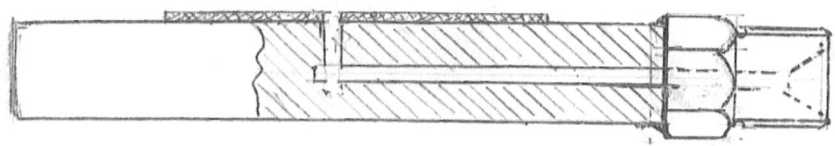

FIG. 5 : PLATE DEVICE (CROSS SEC. AT CEMTRE)

FIG. 6 MULTI-PORT DEVICE FOR PRESSURE TEST.

SPECIMEN: PRESSURE TEST CERTIFICATE FOR SINGLE ITEM

(Type on your company's letter head)

PRESSURE TEST CERTIFICATE

No.(Item Id)   Date: dd/mm/yyyy

      P.O. No. ABCD78910   PO Date: --/--/----
      Specification: (Enter Spec. No/Rev. No/Var. No.)
      Product: (Name of Product(s)
      Order Quantity (NOs):

This is to certify that x % of order quantity was subjected to an internal/external pressure of (999) Kg/Cm² for 20 minutes each.

Result: No leakage or deformation was observed in any of the tested items.

Signature and stamp   Signature and stamp
Of Inspecting Agency   of Supplier/Vendor

SPECIMEN : PRESSURE TEST CERTIFICATE FOR 2 OR MORE ITEMS OF THE SAME SPEC.

(Type on your company's letter head)

PRESSURE TEST CERTIFICATE

No.(Item Id)   Date: dd/mm/yyyy

P.O. No. ABCD78910   PO Date: --/--/----
Specification: (Enter Spec. No/Rev. No/Var. No.)
Product: (Name of Product(s)
Order Quantity (NOs):

This is to certify that x % of order quantity of each item was subjected to an internal/ (external) pressure shown against them for 20 minutes.

PO S.No.	Item Code	Product Name/ Designation	Order Qty.(NOs)	Quantity Tested	Test Press. Kg/Cm²
10					
20					
30					

Result: No leakage or deformation was observed in any of the tested items.

Signature and Seal   Signature and Seal
Of Inspecting Agency   of Supplier/Vendor

SPECIMEN OF GUARANTEE CERTFICATE

Company's Letterhead

Certificate No. (ID No. of 1st item)   Date: dd/mm/yyyy

This is to certify that the following materials supplied against Purchase Order No. "ABC12345" dated dd/mm/yyyy are strictly as per the Specification in the Purchase Order.

If any defect or discrepancy is found and reported within a period of 12/18 months from the date of commissioning/supply, we undertake to replace the defective material free of cost.

PO S.No.	Description	Item Code	ID Mark (Logo+ ID No.)	Quantity (Numbers)
10				
20				
30				

To   For (Your CO's Name)
Customer's name and brief Address

(Name and Signature of Authorized Signatory)